Communications in Computer and Information Science 2976

Series Editors

Gang Li, *School of Information Technology, Deakin University, Burwood, VIC, Australia*
Joaquim Filipe, *Polytechnic Institute of Setúbal, Setúbal, Portugal*
Zhiwei Xu, *Chinese Academy of Sciences, Beijing, China*

AF790230

Rationale
The CCIS series is devoted to the publication of proceedings of computer science conferences. Its aim is to efficiently disseminate original research results in informatics in printed and electronic form. While the focus is on publication of peer-reviewed full papers presenting mature work, inclusion of reviewed short papers reporting on work in progress is welcome, too. Besides globally relevant meetings with internationally representative program committees guaranteeing a strict peer-reviewing and paper selection process, conferences run by societies or of high regional or national relevance are also considered for publication.

Topics
The topical scope of CCIS spans the entire spectrum of informatics ranging from foundational topics in the theory of computing to information and communications science and technology and a broad variety of interdisciplinary application fields.

Information for Volume Editors and Authors
Publication in CCIS is free of charge. No royalties are paid, however, we offer registered conference participants temporary free access to the online version of the conference proceedings on SpringerLink (http://link.springer.com) by means of an http referrer from the conference website and/or a number of complimentary printed copies, as specified in the official acceptance email of the event.

CCIS proceedings can be published in time for distribution at conferences or as post-proceedings, and delivered in the form of printed books and/or electronically as USBs and/or e-content licenses for accessing proceedings at SpringerLink. Furthermore, CCIS proceedings are included in the CCIS electronic book series hosted in the SpringerLink digital library at http://link.springer.com/bookseries/7899. Conferences publishing in CCIS are allowed to use our online conference service (Meteor) for managing the whole proceedings lifecycle (from submission and reviewing to preparing for publication) free of charge.

Publication process
The language of publication is exclusively English. Authors publishing in CCIS have to sign the Springer CCIS copyright transfer form, however, they are free to use their material published in CCIS for substantially changed, more elaborate subsequent publications elsewhere. For the preparation of the camera-ready papers/files, authors have to strictly adhere to the Springer CCIS Authors' Instructions and are strongly encouraged to use the CCIS LaTeX style files or templates.

Abstracting/Indexing
CCIS is abstracted/indexed in DBLP, Google Scholar, EI-Compendex, Mathematical Reviews, SCImago, Scopus. CCIS volumes are also submitted for the inclusion in ISI Proceedings.

How to start

To start the evaluation of your proposal for inclusion in the CCIS series, please send an e-mail to ccis@springer.com

Karl Mason · Usman Haider ·
Marcos Eduardo Cruz Victorio
Editors

Intelligent Agents in Science and Engineering

International Workshop, IASE 2025, Held in Conjunction with ECAI 2025
Bologna, Italy, October 26, 2025
Proceedings

Springer

Editors
Karl Mason
University of Galway
Galway, Ireland

Usman Haider
University of Galway
Galway, Ireland

Marcos Eduardo Cruz Victorio
University of Galway
Galway, Ireland

ISSN 1865-0929 ISSN 1865-0937 (electronic)
Communications in Computer and Information Science
ISBN 978-3-032-28262-0 ISBN 978-3-032-28263-7 (eBook)
https://doi.org/10.1007/978-3-032-28263-7

© The Editor(s) (if applicable) and The Author(s), under exclusive license
to Springer Nature Switzerland AG 2026

This work is subject to copyright. All rights are solely and exclusively licensed by the Publisher, whether the whole or part of the material is concerned, specifically the rights of translation, reprinting, reuse of illustrations, recitation, broadcasting, reproduction on microfilms or in any other physical way, and transmission or information storage and retrieval, electronic adaptation, computer software, or by similar or dissimilar methodology now known or hereafter developed.
The use of general descriptive names, registered names, trademarks, service marks, etc. in this publication does not imply, even in the absence of a specific statement, that such names are exempt from the relevant protective laws and regulations and therefore free for general use.
The publisher, the authors and the editors are safe to assume that the advice and information in this book are believed to be true and accurate at the date of publication. Neither the publisher nor the authors or the editors give a warranty, expressed or implied, with respect to the material contained herein or for any errors or omissions that may have been made. The publisher remains neutral with regard to jurisdictional claims in published maps and institutional affiliations.

This Springer imprint is published by the registered company Springer Nature Switzerland AG
The registered company address is: Gewerbestrasse 11, 6330 Cham, Switzerland

If disposing of this product, please recycle the paper.

Preface

The Workshop on Intelligent Agents in Science and Engineering (IASE 2025) was held in conjunction with the 28th European Conference on Artificial Intelligence (ECAI 2025) in Bologna, Italy, in October 2025.

The objective of this multidisciplinary workshop was to bring together researchers and practitioners to discuss methodological, technical, organizational, and environmental aspects of intelligent agents applied across scientific and engineering domains. In recent years, intelligent agents and multi-agent systems have gained significant attention due to their strong connections with artificial intelligence, machine learning, computational modelling, and distributed systems.

Agents and multi-agent systems have been successfully applied to a wide range of challenges, including modeling and simulation, decision-making, optimization, control systems, robotics, renewable energy systems, smart grids, biomedical engineering, manufacturing, and cloud computing. The workshop provided a forum for presenting advances in agent-based research and exploring practical applications in academia and industry.

IASE 2025 received 7 submissions. All papers underwent a single-blind peer review process and were evaluated by at least two members of the Programme Committee based on relevance, originality, technical soundness, clarity, reproducibility, and overall contribution. Of these submissions, 6 papers were accepted for publication, yielding an acceptance rate of 85.7%. In addition, 4 invited contributions were included in the workshop program.

The workshop featured three keynote presentations delivered by distinguished researchers in the field of intelligent agents and artificial intelligence. Accepted papers were presented as oral or combined poster/spotlight presentations; this allocation was randomized and did not reflect the scientific quality of the contributions.

We would like to sincerely thank the members of the Program Committee for their careful reviews and constructive feedback. We also express our gratitude to the keynote speakers and all authors for their valuable contributions. Finally, we thank the organisers of ECAI 2025 for hosting the workshop in the historic city of Bologna.

March 2026

Karl Mason
Usman Haider
Marcos Eduardo Cruz Victorio

Preface

Organization

Workshop Organizers

Karl Mason	University of Galway, Ireland
Usman Haider	University of Galway, Ireland
Marcos Eduardo Cruz Victorio	University of Galway, Ireland

Program Committee

Abril Corona Figueroa	Durham University, UK
Burak Karaduman	University of Antwerp, Belgium
Fatima Khalid	GIK Institute of Engineering Sciences and Technology, Pakistan
Hossein Yarahmadi	University of Antwerp, Belgium
Ivan Yeniseysk Alba Corpus	National Technological Institute of Mexico, Mexico
Juan Rodolfo Alvarez Padilla	Carnegie Mellon University, USA
Julia Diaz-Escobar	National Technological Institute of Mexico, Mexico
Krzysztof Kluza	AGH University of Science and Technology, Poland
Maribel Guerrero Luis	National Technological Institute of Mexico, Mexico
Martha Correa-Delval	Durham University, UK
Mauro Martini	Politecnico di Torino, Italy
Mimi Zhang	Trinity College Dublin, Ireland
Muhammad Hanif	GIK Institute of Engineering Sciences & Technology, Pakistan
Rachael Shaw	Atlantic Technological University, Ireland
Sedar Olmez	Fujitsu Research Europe
Valentina Breschi	Eindhoven University of Technology, The Netherlands
Weiqi Hua	University of Birmingham, UK
Zahid Halim	National Yunlin University of Science and Technology, Taiwan

Contents

Theoretical Foundations and Algorithms

Two-Way Mentor-Mentee Matching Algorithms Under Varying Constraint Complexities

Daniel O'Driscoll(✉) and Jing Hua Ye

Computer Science, Munster Technological University, Bishopstown, Cork City, Republic of Ireland
dodriscoll4@mymtu.ie, jinghua.ye@mtu.ie

Abstract. Mentor-mentee matching is framed as a multi-constraint optimization problem, balancing hard constraints and optimization objectives. Two-sided matching endeavors to find optimal matchings based on the preferences of participants. This study matches the mentors to mentees in a professional development context. We mainly focus on comparatively analyzing the performance of four algorithmic approaches: Euclidean distance-based matching, K-means clustering, Genetic Algorithm (GA), and Deferred Acceptance (DA). Utilizing a dataset of 8,000 potential mentors and mentees with 13 matching features, we evaluated algorithm performance across 100 randomized iterations under varying constraint conditions and two-sided matching directionality. Two-sided preference consideration magnifies performance differences between algorithms, with specialized approaches (GA and DA) substantially outperforming general methods. Two-sided matching approaches generally reduced similarity scores while increasing computational demands.

Keywords: Two-Way Mentor-Mentee Matching · Genetic Algorithm · Constrained Bipartite Matching

1 Introduction

Mentorship plays a vital role in both educational and professional development: "More than 4 in 10 workers who do not have a mentor say they've considered quitting their job in the past three months, compared with just 25% of those who do have a mentor" [3]. A mentor, defined as a knowledgeable and experienced individual who offers guidance and support, can significantly impact a mentee's growth trajectory. Kram and Isabella pointed out that "Mentors provide young adults with career-enhancing functions, such as sponsorship, coaching, facilitating exposure and visibility, and offering challenging work or protection, all of which help the young person establish a role in the organisation, learn the ropes, and prepare for advancement" [4]. A good relationship between a mentor and a mentee can dramatically impact a mentee's future career development; thus, finding a suitable and competent mentor is crucial for mentees. However,

© The Author(s), under exclusive license to Springer Nature Switzerland AG 2026
K. Mason et al. (Eds.): IASE 2025, CCIS 2976, pp. 3–15, 2026.
https://doi.org/10.1007/978-3-032-28263-7_1

the critical process of matching mentors with appropriate mentees remains a significant challenge. This means that in a pool of potential mentors and interested mentees, finding a match that satisfies the criteria of both sides can be of substantial complexity.

There are two approaches to finding an optimal match between a mentor and a mentee: one-sided matching and two-sided matching. Two-sided matching approaches are well-established processes to allocate mentees to mentors based on participants' preferences. Current approaches often rely on manual pairing processes or simple rule-based algorithms; manual pairing can easily miss out some important requirements and is time-consuming; the rule-based algorithms fail to account for the multidimensional factors involved in successful mentoring relationships. Rule-based systems often struggle with complex decision scenarios that involve multiple interrelated variables [2]. These limitations are particularly evident in mentorship matching. Factors of this matching include industry alignment, experience gaps, communication preferences, availability, expertise areas, and personal attributes. When matches are suboptimal, the mentoring relationship suffers, leading to decreased satisfaction, wasted resources, and potentially negative outcomes for both parties [1].

The main objective of this paper is to comparatively investigate the performance of four algorithmic approaches, Euclidean distance-based matching, K-means clustering, Genetic Algorithm (GA), and Deferred Acceptance (DA), to find high-quality matchings that match as many pairs as possible. Additionally, the challenge of maintaining matching quality under increasing constraint loads on two-sided algorithms represented an unexplored dimension of this problem domain.

The general definition of the mentor-mentee matching problem and some relevant existing approaches subject to this problem are reviewed in Sect. 2. The designed algorithmic methodologies are discussed in Sect. 3. The performance of these algorithmic methodologies is assessed and compared in Sect. 4. This paper concludes with a remark and possible future research directions.

2 Background

The process of matching mentors and mentees has evolved significantly from manual pairing to sophisticated algorithmic approaches. This evolution stems from the recognition that effective mentor-mentee relationships yield substantial benefits, including improved academic outcomes, enhanced professional development, and increased retention in educational and organizational contexts. Effective mentor-mentee matching in educational settings has been associated with significant improvements in student retention and academic performance [7]. In professional and academic settings, well-matched mentorship pairs have been linked to higher retention, faster career progression, and improved research outcomes [8]

Designing effective mentor-mentee matching mechanisms is crucial for improving satisfaction and outcomes in organisational settings, particularly when participants have diverse preferences [9]. A study by the Wharton School of Business found that retention rates were significantly higher for both mentees (72%) and mentors (69%) compared to employees who did not participate in mentoring programs (49%). Additionally, mentored employees were promoted five times more often than those not in a mentoring program, and mentors themselves were promoted six times more often [8]. In educational settings, a systematic review by Flores and Estudillo indicated that approximately 65% of mentees reported that the mentor scheme had a positive impact on their decision to continue university for the next semester [7]. Shah et al. identified that the process of matching mentors and mentees remains largely inefficient and often produces suboptimal pairings. Additionally, a modified k-means clustering is a promising approach addressing the issue of the production of suboptimal pairings, but it focuses on a singular algorithmic approach [11]. This presented an opportunity to expand the algorithmic options and develop a comprehensive comparative framework.

Mentor-mentee matching represents a specialised instance of the broader constrained bipartite matching problem, with unique characteristics and challenges. The problem involves finding optimal pairings between two distinct groups (mentors and mentees) while maximising compatibility, satisfying various constraints, and ensuring that both sides are as pleased with the matching as possible [5,6]. The number of mentors does not have to be the same as the number of mentees; hence, a match does not guarantee that each mentee will be matched with a mentor.

Two-sided approaches, by contrast, incorporate preferences from both mentors and mentees, aiming to create mutually satisfactory pairs. Key characteristics include consideration of reciprocal preferences and compatibility, potential for greater relationship stability and longevity, increased computational complexity and data collection requirements, and greater fairness in preference satisfaction across both groups. Two-sided matching in corporate mentorship programs leads to improved relationship satisfaction and longer match duration, although it may involve higher computational costs.

2.1 Algorithmic Approaches

Euclidean Distance-Based Approach. Euclidean distance-based approaches represent one of the most straightforward and widely implemented methods for mentor-mentee matching. This algorithm calculates the distance between mentors and mentees in a multidimensional feature space, with smaller distances indicating greater similarity.

Given mentor and mentee feature vectors m_i and e_j, the Euclidean distance is calculated as:

$$d(m_i, e_j) = \sqrt{\sum_{k=1}^{n} w_k (m_{ik} - e_{jk})^2} \tag{1}$$

where:

- $d(m_i, e_j)$ is the weighted Euclidean distance between mentor i and mentee j

- m_{ik} is the value of the kth feature for mentor i
- e_{jk} is the value of the kth feature for mentee j
- w_k is the weight assigned to feature k, indicating its relative importance
- n is the total number of features considered

Euclidean matching is highlighted: conceptual simplicity and ease of implementation, computational efficiency with $O(nm)$ complexity for n mentors and m mentees, intuitive interpretation of match rationale, and strong performance with appropriately weighted features [12].

K-Means Clustering. K-means clustering represents a natural approach to mentor-mentee matching, particularly when adapted to treat one group (e.g., mentors) as fixed centroids around which the other group (mentees) clusters [10]. For mentor-mentee matching, the standard K-means formulation is adapted by treating the mentors $M = \{m_1, m_2, \ldots, m_n\}$ as fixed centroids, and assigning mentees to the nearest mentor based on feature similarity:

$$\arg\min_{A} \sum_{j=1}^{m} \|e_j - m_{A(j)}\|^2 \tag{2}$$

subject to the capacity constraint:

$$|\{j : A(j) = i\}| \leq c_i \quad \forall i \in \{1, 2, \ldots, n\} \tag{3}$$

where:

- e_j is the feature vector of mentee j
- m_i is the feature vector of mentor i (used as a fixed centroid)
- $A(j)$ is the assignment function mapping each mentee j to a mentor i
- c_i is the maximum capacity (i.e., number of mentees) that mentor i can accept

A modified K-means approach specifically for mentor-mentee matching with their MentorMate platform [10] was built; this platform treats mentors as fixed cluster centroids, incorporates capacity constraints, adds similarity thresholds for valid matches, and implements constraint checking during assignment.

Genetic Algorithm. GAs represent a sophisticated approach to mentor-mentee matching, framing the problem as an evolutionary optimization process. In a GA for mentor-mentee matching, a chromosome A represents a complete matching solution, where:

$$A = [A_1, A_2, \ldots, A_m] \tag{4}$$

Each element A_j indicates which mentor is assigned to mentee j, such that:

$$A_j \in \{1, 2, \ldots, n, -1\} \tag{5}$$

where n is the total number of mentors, m is the total number of mentees, and -1 represents an unassigned state.

A GA for corporate mentorship matching is implemented. A chromosome is encoded as an integer array chromosome representation. A fitness function combines similarity scores and constraint violations. This standard evolutionary mechanism comprises tournament selection, multi-point crossover with constraint repair mechanisms, and random reassignment mutation with constraint verification.

Deferred Acceptance Algorithm. The DA algorithm, introduced by Gale and Shapley, was originally developed to solve the stable marriage problem but has been adapted for mentor-mentee matching. In the DA algorithm, each mentor m_i has a strict preference relation $\succ_{m_i}$ over mentees, and each mentee e_j has a strict preference relation $\succ_{e_j}$ over mentors, where M represents the set of all mentors and E represents the set of all mentees. A matching μ is a function from the set $M \cup E$ into $M \cup E$ such that:

- $\mu(m_i) \subseteq E$ with $|\mu(m_i)| \leq c_i$ for each mentor $m_i \in M$, where c_i represents the capacity of mentor i
- $\mu(e_j) \in M \cup \{e_j\}$ for each mentee $e_j \in E$, where $\mu(e_j) = e_j$ indicates mentee j is unmatched
- $e_j \in \mu(m_i)$ if and only if $\mu(e_j) = m_i$, ensuring consistency in the matching

A modified DA algorithm for corporate mentoring programs was implemented [13]; this enhanced algorithm produced particularly stable matches with high mutual satisfaction.

3 Architecture and Methodology

3.1 Formal Mathematical Formulation of Matching Problem

The mentor-mentee matching problem can be formally defined within the framework of constrained bipartite graph matching. Given:

- A set of mentors $M = \{m_1, m_2, ..., m_n\}$
- A set of mentees $E = \{e_1, e_2, ..., e_m\}$
- A similarity function $S : M \times E \rightarrow [0, 1]$ that assigns a normalised similarity score to each potential pair
- A set of constraints $\Omega = \{\omega_1, \omega_2, ..., \omega_k\}$ defining valid assignments

The objective is to find an assignment $A \subseteq M \times E$ that maximises the overall similarity $\sum_{(m,e)\in A} S(m, e)$ subject to:

Capacity constraint: $\forall m \in M : |\{e|(m, e) \in A\}| \leq \text{max\_mentees\_per\_mentor}$
Validity constraint: $\forall (m, e) \in A : \text{num\_similarities}(m, e) \geq \text{min\_similarities}$
Domain-specific constraint: $\forall (m, e) \in A, \forall \omega_i \in \Omega : \omega_i(m, e) = \text{true}$

3.2 Hard Constraints

There are two key constraints - basic constraints and selected domain-specific constraints. The basic constraints contain two fundamental constraints stating that the relationship between the mentor and the mentee is one-to-one:

max_mentees_per_mentor: Maximum number of mentees a single mentor can be assigned, vice versa, the maximum number of mentors a single mentee can have.
min_similarities: Minimum number of shared characteristics required for a valid match

The selected domain-specific constraints are formulated as below:

industry_match: $\omega_{\text{industry}}(m, e) = \mathbb{1}[m.\text{Industry} = e.\text{Industry}]$
education_level: $\omega_{\text{education}}(m, e) = \mathbb{1}[L(m.\text{Education}) \geq L(e.\text{Education})]$ where L maps education levels to an ordinal scale
position_progression:

$$\omega_{\text{position}}(m, e) = \mathbb{1}[P(m.\text{Experience\_Level}) - P(e.\text{Experience\_Level}) \geq \text{min\_gap}]$$

where P maps positions to an ordinal scale and min_gap is the minimum position level difference required.

expertise_overlap: $\omega_{\text{expertise}}(m, e) = \mathbb{1}[m.\text{Expertise} \cap e.\text{Expertise} \neq \emptyset]$
communication_match: $\omega_{\text{communication}}(m, e) = \mathbb{1}[m.\text{Communication\_Methods} \cap e.\text{Communication\_Methods} \neq \emptyset]$

This approach allowed for a systematic analysis of how increasing constraint complexity affects each algorithm's performance.

3.3 Data Preprocessing Framework

All algorithms utilize a common preprocessing framework to ensure fair comparison. This framework handles the following three important sequential processes:

Feature Encoding: Categorical variables (e.g., industry, language, education, etc.) are encoded using a consistent label encoding mechanism.
Feature Weighting: Domain-specific importance is reflected through feature weighting (see Table 1): $X_{weighted} = W_i \times X_{encoded}$ where W_i represents the importance weight assigned to feature i and $X_{encoded}$ represents the features resulted from the feature encoding process.
Data Normalization: Feature scales are standardized, $X_{norm} = \frac{X - min(X)}{max(X) - min(X)}$ where X_{norm} represents the normalized feature, to prevent bias toward a particular dimension.

Table 1. Feature Importance Weights

Feature	Weight
Position	1.5
Experience Level	1.2
Primary Language	1.3
Industry	1.4
Availability	1.0
Communication Methods	1.0
Mentoring Preferences	1.5
Education	1.2

3.4 Algorithmic Core

Euclidean Approach: Calculates weighted similarity between each mentee and mentor using distance in feature space:

$$\text{similarity}(e_j, m_i) = 1 - \frac{d(e_j, m_i)}{\text{max\_distance}} \tag{6}$$

Mentees are assigned to their most similar available mentor, respecting capacity constraints.

K-means Approach: Treats mentors as fixed cluster centroids with mentees assigned to optimal clusters. Modified K-means incorporates mentor capacity constraints, similarity thresholds, and constraint checking during assignment.

GA Approach: Models matching solutions as chromosomes where each gene represents a mentee-mentor assignment. Fitness evaluation combines similarity scores and constraint satisfaction through tournament selection, multi-point crossover, and mutation operations.

DA Approach: Implements preference-driven matching with iterative proposal rounds from unmatched mentees to preferred mentors, tentative acceptance based on mentor preference and capacity, and constraint verification during acceptance.

4 Experiments and Analytical Results

4.1 Dataset

This study utilized a comprehensive mentor-mentee dataset consisting of 8,000 profiles (4,000 mentors and 4,000 mentees) with 13 profile attributes per record, including position, experience level, expertise, primary language, secondary languages, gender, industry, education, years of experience, availability, location, mentoring preferences, and communication methods. The default test sample sizes were 500 mentors and 1,000 mentees, with scaling tests ranging from small (500/1000) to maximum (4000/4000) dataset sizes.

Table 2. The general experimental settings across all algorithms

Parameter	Value
Mentor Sample Size	500 (baseline)
Mentee Sample Size	1000 (baseline)
Iterations Per Setting	100 (for randomised testing)
Basic Constraints	max_mentees_per_mentor: 2, min_similarities: 2
Parallel Processes	4 (for parallelised execution)

Table 3. The parameter settings for the GA approach and the k-means approach

Algorithmic Parameter	Value
GA - Generations	20
GA - Population Size	100
GA - Crossover Rate	0.9
GA - Mutation Rate	0.01
GA - Tournament Size	5
K-means - Clusters	10

4.2 General Experimental Settings

The general experimental settings across all developed algorithms are presented in Table 2. The parameter settings for the GA approach and the k-means approach are presented in Table 3. The experiments were conducted with a standard Dell Precision 3660 with an NVIDIA RTX A2000 GPU (12GB VRAM), Intel Core i7-12700 processor, and 32 GB of system RAM. All algorithms were built with the standard Python library. The experiments were conducted with the systematic variation of parameters:

- Constraint complexity (2, 4, 6, 10, 20 constraints)
- Dataset size (from 500 mentors/1000 mentees to 4000/4000)
- Algorithmic-specific parameters (e.g., generations for GA)

The similarity measurement formed the foundation of the matching quality assessment.

4.3 Two-Sided Randomised Constraint Analysis

To evaluate how mutual preference consideration affects algorithm performance under increasing constraint complexity, we conducted two-sided implementation testing across all four algorithms. This approach reveals important differences in how algorithms handle two-sided preferences while managing constraint interactions across 100 iterations per configuration.

Table 4. Two-Sided Table: Summary of Matching Algorithms Across All Constraint Levels

Algorithm	Avg. Validity (%)	Avg. Similarity (%)	Avg. Execution Time (s)
Euclidean	94.19	77.75	72.02
K-Means	40.39	19.3	134.47
GA	95.26	91.78	648.5
DA	93.58	87.28	45.8

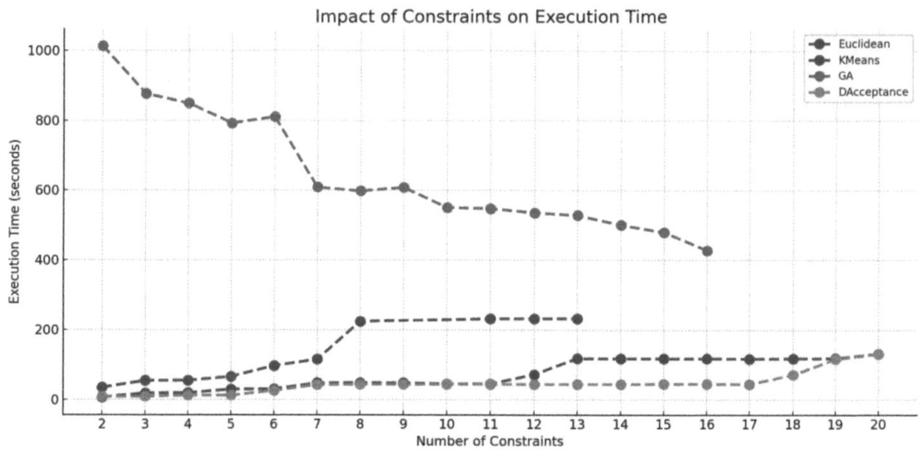

Fig. 1. Execution Time vs Randomised Constraints

Table 4 presents a comprehensive comparison of the four algorithms in their two-sided implementations. In contrast to the one-sided results, the two-sided implementations show more pronounced differences between algorithms. The GA achieves exceptional performance with the highest validity (95.26%) and similarity scores (91.78%), demonstrating that its flexible optimization framework effectively balances two-sided preferences, albeit at substantially higher computational cost (648.5 s). DA, designed specifically for two-sided matching problems, achieves comparable validity (93.58%) and impressive similarity scores (87.28%) with the fastest execution time (45.8 s), confirming its theoretical advantages for two-sided scenarios. Euclidean distance maintains good validity (94.19%) but with lower similarity scores (77.75%), suggesting its simplified approach struggles to fully optimize mutual preferences. Most notably, K-means performance deteriorates dramatically in the two-sided implementation, with validity dropping to 40.39% and similarity plummeting to 19.3%, indicating fundamental limitations in handling dual preference structures.

Figure 1 illustrates how two-sided preference consideration affects execution time across increasing constraint complexity. The most notable observation is the GA's consistently high computational demand, starting at approximately

1000 s with minimal constraints and gradually decreasing to around 420 s as the constraint count increases. This counterintuitive pattern differs significantly from one-sided implementation and suggests that two-sided preference consideration creates a more complex initial search space that becomes more manageable as constraints narrow the feasible region. The K-means algorithm experiences a dramatic execution time increase (from approximately 50 to 220 s) when constraint count reaches 7–8, indicating a critical complexity threshold where the algorithm struggles to reconcile two-sided preferences with multiple constraints. In contrast, both Euclidean and DA algorithms maintain relatively stable execution profiles until high constraint levels, with DA showing particular efficiency (under 50 s) until constraint level 18, where it experiences a significant increase. This pattern demonstrates that algorithms specifically designed for two-sided matching (DA) handle preference-constrained spaces more efficiently than general-purpose approaches.

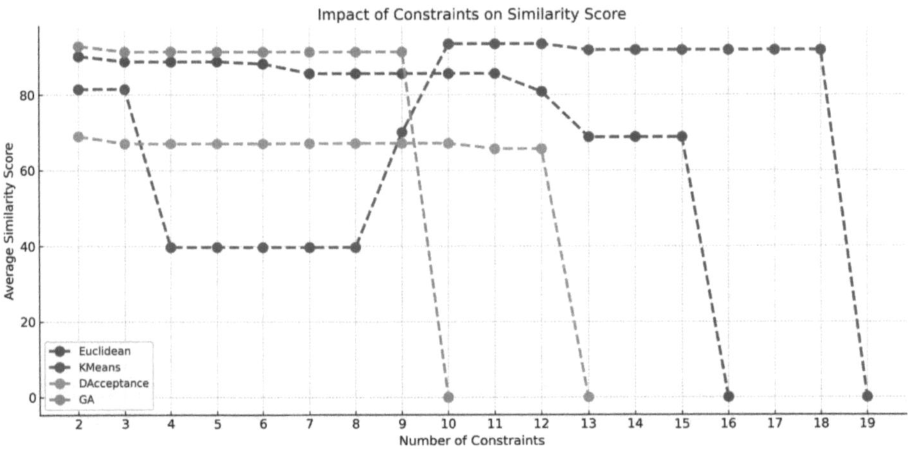

Fig. 2. Similarity Score vs Randomised Constraints

The similarity score analysis (see Fig. 2) reveals fundamental differences in how algorithms maintain match quality when balancing two-sided preferences under increasing constraints. Unlike the one-sided results, we observe greater differentiation between algorithms. GA and DA maintain exceptionally high similarity scores (approximately 90–95%) across most constraint levels, with GA slightly outperforming DA until the highest constraint levels. This demonstrates that both algorithms effectively optimize for mutual preferences even under complex constraints. The Euclidean approach performs adequately with scores around 85% initially, dropping to approximately 78% after constraint level 12, showing moderate capacity to balance two-sided preferences. Most significantly, K-means exhibits dramatically lower similarity scores (approximately 25%) even with minimal constraints, and completely fails after constraint level

12. This substantial degradation compared to its one-sided performance confirms that the clustering approach fundamentally struggles to handle two-sided preference structures when constraints are applied. The universal drop to zero similarity at constraint level 20 indicates that no algorithm can satisfy this extreme constraint combination.

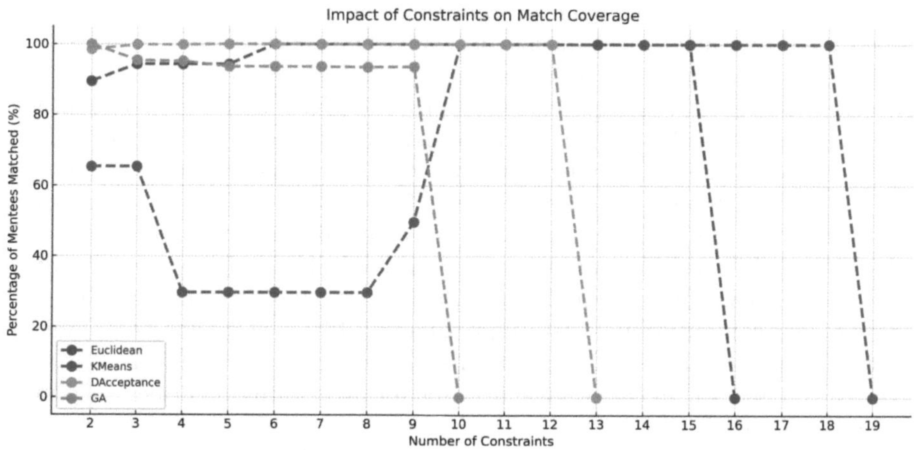

Fig. 3. Validity Percentage vs Randomised Constraints

The validity analysis (see Fig. 3) completes our understanding by showing how constraint complexity affects the percentage of valid matches in two-sided implementations. The most striking difference from one-sided results is the significantly poorer performance of K-means, which fluctuates between 30–60% validity before completely failing after constraint level 11. This suggests that introducing two-sided preferences substantially amplifies the algorithm's sensitivity to constraints, creating compatibility challenges that exceed its problem-solving capacity. In contrast, Euclidean, GA, and DA all demonstrate excellent robustness, maintaining near-perfect validity (approximately 100%) across most constraint levels until the universal breaking point at level 20. This pattern indicates that these algorithms can effectively satisfy constraints while balancing two-sided preferences, with DA showing particular stability, achieving perfect 100% validity across nearly all constraint levels despite relatively fast execution times. The consistent validity drop across all algorithms at constraint level 20 confirms that this combination (adding position progression to an already heavily constrained problem) creates an infeasible matching scenario.

5 Conclusion and Future Work

The optimization of mentor-mentee matching represents an important challenge with significant implications for educational and professional development. By approaching this challenge as a formal optimization problem and systematically evaluating different algorithmic approaches, this study contributes to both theoretical understanding and practical implementation of effective matching systems.

This research comprehensively evaluated four distinct two-sided mentor-mentee matching algorithms under varying conditions of constraint complexity. The analysis revealed significant differences in performance, scalability, and resilience across Euclidean, K-means, GA, and DA approaches.

Two-sided preference consideration magnifies performance differences between algorithms, with specialized approaches (GA and DA) substantially outperforming general methods. K-means experiences catastrophic performance degradation in two-sided implementation, with validity dropping by approximately 28% points and similarity by over 50% points compared to one-sided implementation. DA achieves the most favorable balance of execution efficiency and match quality, demonstrating its theoretical advantages for two-sided matching problems. GA achieves marginally higher validity and similarity but at significantly higher computational cost (approximately 14× slower than DA), limiting its practicality for large-scale applications. The position_progression constraint (level 20) creates an infeasible scenario for all algorithms regardless of implementation approach.

These findings highlight the critical importance of algorithm selection when implementing two-sided mentor-mentee matching systems. Systems requiring mutual preference consideration should prioritize specialized algorithms like DA that balance efficiency with quality, or GA when computational resources permit, and match optimization is paramount.

As mentorship continues to play a vital role in knowledge transfer and skill development across various domains, the insights gained from this research provide a foundation for more effective and efficient matching systems that maximize the benefits of mentorship relationships. Future work will expand these findings by incorporating additional algorithms, exploring more complex constraint scenarios, and validating results with real-world implementation data.

Acknowledgments. We would like to thank the Department of Computer Science at Munster Technological University for supporting this research.

References

1. Cook, S.: Mentorship Red Flags: Signs Your Mentoring Program at Work Isn't Working (2023). https://www.mentorcliq.com/blog/mentorship-programs-that-dont-work
2. Chu, D.: Overcoming the limitations of rule-based systems. In: Secoda Blog (2024). https://www.secoda.co/blog/overcoming-the-limitations-of-rule-based-systems
3. Wronski, L., Cohen, J.: Nine in 10 workers who have a career mentor say they are happy in their jobs. CNBC (2019). https://www.cnbc.com/2019/07/16/nine-in-10-workers-who-have-a-mentor-say-they-are-happy-in-their-jobs.html

4. Kram, K.E., Isabella, L.A.: Mentoring alternatives: the role of peer relationships in career development. Acad. Manag. J. **28**(1), 110–132 (1985). https://www.bu.edu/sph/files/2012/01/Kram_Mentoring-Alternatives.pdf
5. Wang, H., Ma, C., Zhou, L.: Bi-matching: a constrained bipartite matching approach with applications in educational recommendation systems. In: IEEE International Conference on Data Mining (ICDM), pp. 627–636 (2022). https://ieeexplore.ieee.org/document/9864532
6. Gross, J., Koutsopoulos, I.: The k-constrained bipartite matching problem: approximation algorithms and applications to wireless networks. In: Proceedings of IEEE INFOCOM, pp. 1–5 (2010). https://www.jamesgross.org/wp-content/uploads/2016/01/05462027.pdf
7. Le, H.-G., Sok, S., Heng, K.: The benefits of peer mentoring in higher education: findings from a systematic review. J. Learn. Dev. High. Educ. **31** (2024). https://doi.org/10.47408/jldhe.vi31.1159
8. Winstanely, G.: Mentoring Statistics You Need to Know - 2025 (2025). https://mentorloop.com/blog/mentoring-statistics/
9. Haas, C., Hall, M.: Two-sided matching for mentor-mentee allocations-algorithms and manipulation strategies. PLoS One **14**(3), e0213323 (2019). https://doi.org/10.1371/journal.pone.0213323
10. Shah, S., Timorabadi, H., Dasadia, S., Syed, S., Muhammad, D.: Mentormate: a platform to empower students' path to success through tailored mentor-mentee pairings. In: ASEE Annual Conference & Exposition, pp. 1–10 (2023)
11. Shah, S., Timorabadi, H.S., Dasadia, S., Khatib Syed, S., Muhammad, D.: Board 47: a mentor-mentee matching algorithm to automate process of finding an ideal mentor for students. In: ASEE Annual Conference & Exposition, Portland, Oregon (2024). https://doi.org/10.18260/1-2-47043
12. Rodriguez-Doñate, M., Hernandez, R., Gonzalez, E.: Euclidean distance-based matching for medical residency mentorship programs. J. Med. Educ. Technol. **32**(4), 387–401 (2021). https://doi.org/10.1080/10872981.2021.1968432
13. Williams, C., Zhang, R.: Two-sided stable matching in mentorship programs: comparing deferred acceptance to alternative approaches. In: IEEE Symposium on Computational Intelligence, pp. 224–231 (2022). https://ieeexplore.ieee.org/document/9726842

Refutation of Spectral Graph Theory Conjectures with Search Algorithms

Milo Roucairol and Tristan Cazenave(✉)

LAMSADE, Université Paris Dauphine - PSL, Place du Maréchal de Lattre de Tassigny, 75016 Paris, France
tristan.cazenave@dauphine.fr

Abstract. We are interested in the automatic refutation of spectral graph theory conjectures. Most existing works address this problem either by exhaustively generating graphs of limited size or by using deep reinforcement learning. Exhaustive generation is limited by the size of the generated graphs, and deep reinforcement learning takes hours or days to refute a conjecture. We propose to use search algorithms to address these shortcomings to find potentially large counter-examples to spectral graph theory conjectures in seconds. We apply a wide range of search algorithms to a selection of conjectures from Graffiti. Out of 13 already refuted conjectures from Graffiti, our algorithms are able to refute 12 in seconds. We also refute conjecture 197 from Graffiti, which was open until now.

Keywords: Monte Carlo Search · Spectral Graph Theory · Conjecture · Refutation

1 Introduction

Monte Carlo search algorithms have proven to be powerful as game-playing agents, with recent successes like AlphaGo [14]. These algorithms have the advantage of only needing an evaluation function for the final state of the space they explore.

Graph conjectures are propositions on graph classes (any graph, trees, $K_n - free...$) that are suspected to be true and are awaiting proof or a refutation. They lend themselves well to computer-assisted proofs, as finding a counterexample can be tedious to do manually. Spectral graph conjectures are appropriate for automated refutation because the property can often be directly turned into the evaluation function that can take many different values. Thanks to software like Auto-GraphiX [9] and Graffiti [5], there are plenty of such conjectures.

Adam Zsolt Wagner showed that one could find explicit counterexamples using deep reinforcement learning of a policy with the deep cross-entropy method [16]. Wagner was able to refute three conjectures from Graffiti. Wagner's method trained a neural network with reinforcement learning for each conjecture, learning to build graphs tailored to the refutation of each conjecture. The refutation

© The Author(s), under exclusive license to Springer Nature Switzerland AG 2026
K. Mason et al. (Eds.): IASE 2025, CCIS 2976, pp. 16–27, 2026.
https://doi.org/10.1007/978-3-032-28263-7_2

of a conjecture with Wagner's method takes hours or days. Using search algorithms, it was possible to refute conjectures 2.1 and 2.4 from the Wagner paper in one second and conjecture 2.3 in 291 s [12].

A usual way to find counterexamples to graph theory conjectures is to generate all the graphs smaller than a given size and to verify the conjecture for all these small graphs [1]. This exhaustive generation limits the size of the graphs that can be tested. On the contrary, our search algorithms can rapidly generate larger refutation graphs.

In this paper, the search algorithms will play the game of refuting conjectures by building counterexamples edge by edge. We build on our previous paper [12] by testing many more conjectures and by comparing more search algorithms.

First, we will present the refutation of graph theory conjectures, then the different algorithms we use to explore the problem space, after that, the procedure we use to build graphs and the game rules, and finally, we expose our new results on multiple different conjectures.

2 Refutation of Graph Theory Conjectures

2.1 Graph Theory Conjectures

Refuting Graph theory conjectures can be a hard task to do manually. Imagining a large number of graphs and computing invariant or NP-hard problem values can be tedious and often results in a waste of time. Computers are designed to help with these score computations. The goal of automated conjecture refutation is to automate the exploration as well.

Graph theory conjecture can concern many different properties of graphs: existence, topology, flow, connectivity, cycle, minors, spectral... Here, we are interested in the spectral graph theory conjectures, necessitating matrix calculations only. The spectrum of a matrix is the set of its eigenvalues. Spectrum-related invariants on different types of graph-related matrices (adjacency, distance, Laplacian...) are the focus of spectral graph theory conjectures.

We propose to compare search algorithms for the refutation of various spectral graph theory conjectures. In [12], the conjectures from [16] were refuted hundreds of times faster, and the performances of the methods were compared. This paper experiments with a more diverse set of search algorithms, and we address all the remaining conjectures from Graffiti that do not involve solving an NP-hard problem. Graffiti is a well-known software that was used to generate a thousand spectral graph theory conjectures. It was the target of numerous works and discussions, sometimes involving some of the greatest graph theorists such as Paul Erdos, Laszlo Lovasz, Noga Alon, Noam Nisan. Conjectures from Graffiti were the subject of entire articles, with graph theorists dedicating weeks or months to them. Our program aims at refuting some of them in a much more efficient manner.

We decided to focus on this subset of Graffiti conjectures to avoid dependence on other algorithms and to reduce the number of conjectures.

2.2 Algorithms Used to Refute Graph Conjectures

The program by Wagner trained a deep neural network with cross-entropy to output a policy. The network is used to learn a policy from a state. The policy is used to generate a batch of graphs, it then evaluates the best graphs from the batch and trains the neural network using the graphs' scores. Batches are successively evaluated until the refutation of the conjecture.

Software used for automated conjecture generation, like Graffiti or AutoGraphiX, also has its ways of automatically refuting conjectures to verify the easily refutable ones. Graffiti and its conjectures are detailed in Aouchiche and Hansen's survey [1]. It generates many conjectures in the form of inequalities between invariants. It then tests these conjectures on a database of graphs and discards the falsified ones. Then it checks if the inequalities are not implied by already known theorems and conjectures. If a conjecture passes these tests, it is proposed to scientists. "Written on the wall" collects almost 1000 conjectures from Graffiti, along with the discussions of many scientists. Graffiti's conjectures connect various topics in spectral graph theory, such as eigenvalues, adjacency, distance, Laplacian, and gravity matrices of graphs. These conjectures are the subjects of dozens of articles.

3 Materials and Methods

We decided to expand the experiments from [12] by using the same algorithms and adding two widely used Monte Carlo Search algorithms, another recent one, and another baseline simple algorithm on more conjectures produced by Graffiti [1]. The search model used by the algorithms to build the graphs is the same as in [12]: starting from a single node, new nodes and edges are added to the graph until it reaches a target size. These additions of nodes and edges are the edges of the search tree.

- NMCS [3] uses nested levels of search with random search at the base level. It is already used in [12].
- NRPA [11] learns a playout policy with nested levels of best sequences. At the lowest level, it makes playouts with the learned policy. It is already used in [12].
- GBFS is a simple greedy algorithm opening the best state from a list, evaluating the children of this state, and inserting these children in the list according to their evaluation. It is already used in [12].
- BEAM search, another baseline greedy search heuristic, keeps the *width* best states after each step of expanding and going down the search tree.
- Upper Confidence bounds applied to Trees (UCT) as described by Kocsis and Szepesvári [10], which is the most widely used MCTS algorithm, and close to PUCT used in outstanding works such as Deepmind's Alphazero [15].
- Rapid action value estimation (RAVE), a MCTS algorithm inspired by UCT and proposed by Siver and Gelly [8], then generalized as GRAVE by Cazenave [4] using a threshold on the number of playouts.

- Lazy Nested Monte Carlo Search (LNMCS), a variant of NMCS proposed by Roucairol and Cazenave [13] addresses a shortcoming of the NMCS. Before launching a costly lower-level LNMCS on a child state, it is evaluated with playouts and pruned if the evaluation is not satisfying enough, this way it avoids sinking the time budget in fruitless subtrees and allows the LNMCS to use higher starting levels than the NMCS.

These conjectures (in Table 1) were selected because they did not require NP-hard problems to be solved to compute the score, thus lending themselves well to Monte Carlo Search and other optimization methods. All of the fitting conjectures from Aouchiche and Hansen's survey [1] are included in our experiments. It is possible to tackle conjectures needing an NP-hard problem to be solved (like coloration) up to sizes 25-30, we might come back to such conjectures in future works.

4 Results

The experiments were run with Rust 1.59 on an Intel Core i5-6600K 3.50 GHz CPU, using a single core (but parallel processing is readily available).

Our positive results in Table 1 were obtained by launching each algorithm on each conjecture with a time budget of 15 min. In the Graffiti column, R means the conjecture was already refuted, and O means the conjecture is still open.

- The NMCS used a level of 3.
- The LNMCS used a level of 4, 3 playouts per evaluation, and a ratio of 0.8.
- The UCT used a constant of 1.
- The GRAVE and RAVE used a reference of 5.
- The BEAM search used a width of 10.

5 Discussion

As shown in Table 1, we were unable to refute any of the open conjectures except one (either 197 or 322). This means either that our algorithms are not strong enough, or that these conjectures are true. However, with the exception of Graffiti 140, most of the already refuted conjectures we tried were refuted by at least one algorithm among GRAVE, RAVE, NMCS, LNMCS, NRPA, GBFS, and UCT. Graffiti 140 refutation seems to be lost, 140 is said to be refuted by another work by Favaron, which we are unable to find (Fig. 1).

Considering nonterminal states for evaluation helped immensely with the harder Graffiti 137, 139, 189, 289, and 301. Graffiti 289's repeating pattern may explain why the NRPA succeeded where most other algorithms failed. The times, when displayed in hundreds of seconds, are approximations; only GBFS and BEAM are deterministic, so their times are exact. In addition, NMCS, UCT,

Table 1. Times in seconds to obtain a refutation by applying every algorithm on selected Graffiti conjectures. Graffiti: R means the conjecture was already refuted (as reported in [1]), O means the conjecture is open to be proved or refuted. Size: If there is a "+", it is the size at which we start to find counterexamples with our algorithms. Otherwise, it is the size at which we built the graphs. Graph type: It is the types of graphs we used in our searches. "any" means there was no restriction on the graph built, $K3 - free$ means there is no clique of size 3, "girth $\geq$ 5" means the shortest cycle must be longer than 5, "tree" means the graph must be a tree. We tried both generating trees and then any graph on multiple conjectures. A 0 in the time section means that the conjecture was refuted in less than a second.

Graffiti	Size	Graph type	NMCS	LNMCS	NRPA	UCT	GBFS	BEAM	GRAVE	RAVE
166 R	5+	any	0	0	0	0	0	0	0	0
321 R	4+	$K3 - free$	0	0	0	0	0	0	0	0
189 R	5+	any	0	0	0	2	10	4	2	–
289 R	20+	girth $\geq$ 5	600	600	200	–	6	102	–	–
302 R	6+	tree	0	0	0	0	0	0	0	0
715 R	16+	tree	0	0	0	0	1	0	0	0
29 R	7+	any	2	2	10	5	0	–	2	1
30 R	12+	any	0	0	0	0	311	–	0	0
301 R	14+	tree	2	2	4	–	0	0	7	2
137 R	67+	any	–	–	–	–	513	–	–	–
139 R	50+	any	–	–	–	–	36	–	–	–
711 R	5+	any	0	0	4	0	0	0	0	0
197 O	17+	$K3 - free$	30	30	5	–	0	4	–	–
140 R	50	any & tree	–	–	–	–	–	–	–	–
290 O	50	girth $\geq$ 5	–	–	–	–	–	–	–	–
284 O	50	girth $\geq$ 5	–	–	–	–	–	–	–	–
195 O	50	any	–	–	–	–	–	–	–	–
21 O	50	any & tree	–	–	–	–	–	–	–	–
39 O	50	any & tree	–	–	–	–	–	–	–	–
143 O	100	any & tree	–	–	–	–	–	–	–	–
154 O	50	any & tree	–	–	–	–	–	–	–	–
20 O	50	any & tree	–	–	–	–	–	–	–	–
40 O	50	any & tree	–	–	–	–	–	–	–	–
254 O	50	any & tree	–	–	–	–	–	–	–	–
262 O	50	any & tree	–	–	–	–	–	–	–	–
712 O	50	any & tree	–	–	–	–	–	–	–	–
198 O	50	any	–	–	–	–	–	–	–	–
714 O	100	any	–	–	–	–	–	–	–	–
219 O	50	$K3 - free$	–	–	–	–	–	–	–	–
322 O	50	$K3 - free$	–	–	–	–	–	–	–	–
292 O	50	girth $\geq$ 5	–	–	–	–	–	–	–	–
295 O	50	girth $\geq$ 5	–	–	–	–	–	–	–	–
129 O	50	any	–	–	–	–	–	–	–	–
698 O	50	any	–	–	–	–	–	–	–	–
252 O	50	any	–	–	–	–	–	–	–	–

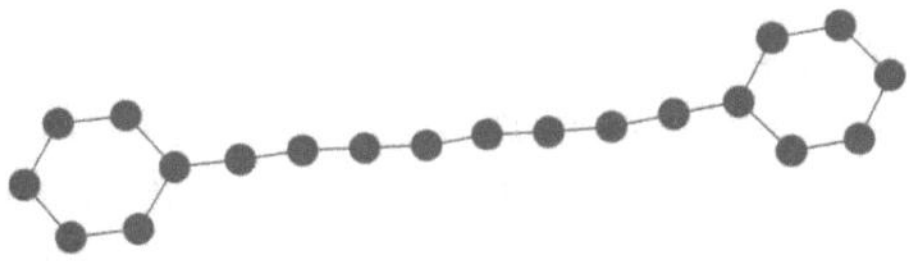

Fig. 1. A counter-example of Graffiti 289 of size 20 (second largest eigenvalue $\leq$ mean of the mean of all adjacent vertex degrees for all nodes)

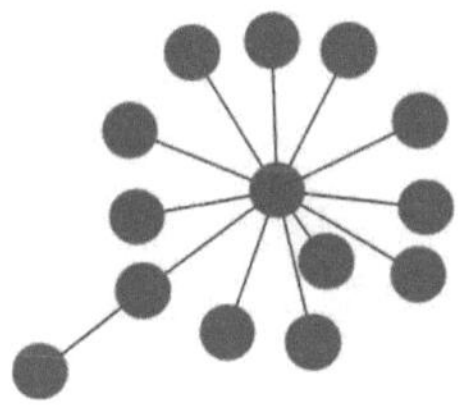

Fig. 2. A counter-example of Graffiti 301 of size 14 (scope of positive eigenvalues $\leq$ harmonic)

GRAVE, and RAVE were very unreliable at finding a counter-example for Graffiti 29, about once every 3 runs, when LNMCS was able to find a counter-example in less than 60 s in 9 out of 10 runs.

For Graffiti 189 and 195, we first searched in the "any graph" search space (instead of the type of graphs the conjecture is about), and then verified if the graph in question complied with the conjecture's definition (Fig. 2).

We think the relative failure of GBFS on Graffiti 30 is partly due to the score function leading the algorithm to open nodes in smaller graphs first, whereas MCTS algorithms try larger graphs immediately via playouts. Beam search requires a width of 80 or more to find a counterexample to this conjecture, so trying larger graphs is not sufficient on its own.

The beam search is very dependent on its width parameter; too high and it will not be able to explore deep in the search tree and find large counterexamples (on Graffiti 137), and too low, and it will not consider inferior intermediate states that would, in fact, lead to a counterexample (on Graffiti 29 and 30). That makes it a less effective choice for refuting the spectral graph conjecture than GBFS. In [12] Graffiti 137 was one of the 4 main conjectures refuted. Unfortunately, an erroneous definition of the Harmonic was used (see Subsect. 5.2). Graffiti 137 is more interesting than we initially thought. The smallest known counterexample featured 101 nodes and required a very specific structure; it was first refuted by Favaron et al. [6] using a mathematical proof rather than a direct counterexample. Here we managed to find a counter-example of size 67 (see Fig. 5) which does not follow the structure described, but closely resembles it, by descending the search space (opening graphs of increasing size, much like conjecture 2) almost directly with GBFS. As mathematical proofs are complex, if the goal was to refute this one conjecture, then our algorithm was most likely faster. But another interesting result is that the class of graphs needed to refute

Graffiti 137 by Favaron et al. was also used to refute two other conjectures (and could have been used to refute Graffiti 139, see Fig. 6). The search algorithms can provide insights to the mathematicians (they can provide insights even if the graph produced does not refute a conjecture as in Wagner's work [16] with conjecture 2.3). The other algorithms did not find a counterexample within a reasonable time (Figs. 3 and 4).

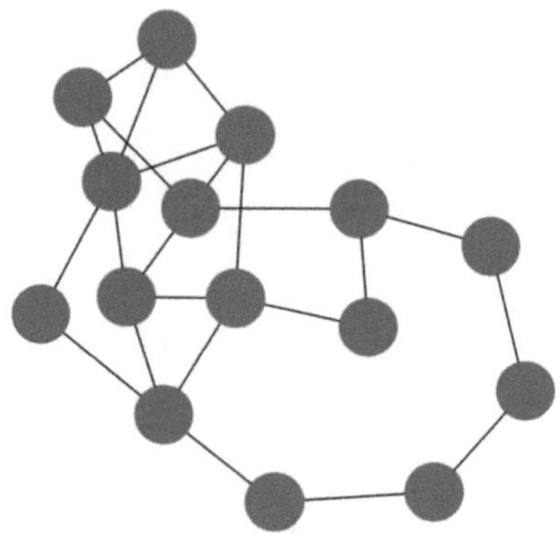

Fig. 3. A counter-example of Graffiti 30 of size 15 (# positive distance eigenvalues $\leq$ sum of temperatures)

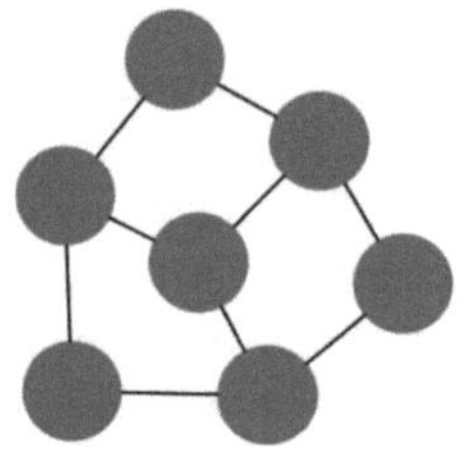

Fig. 4. A counter-example of Graffiti 29 of size 7 (randic index $\leq$ # negative eigenvalues)

5.1 Graffiti 197 Is Refuted

In "Written on the wall", Graffiti 197 is defined as:

Graffiti 197. - 2nd smallest eigenvalue $\leq$ range of eigenvalues of the gravity matrix.

Using the usual definition of the range, and not Aouchiche and Hansen's (see Subsect. 5.2), our algorithms find a counterexample with the cycle of length 17,

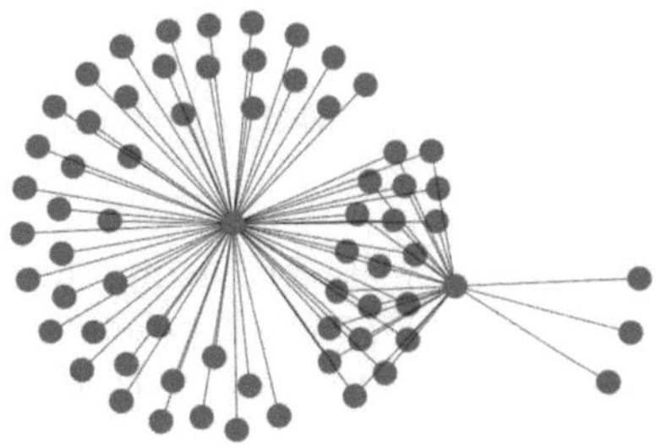

Fig. 5. A counter-example of Graffiti 137 of size 67 (second largest eigenvalue $\leq$ harmonic)

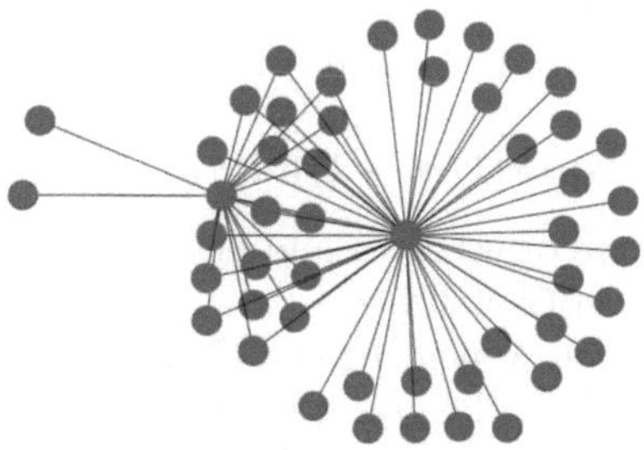

Fig. 6. A counter-example of Graffiti 139 of size 50 (-second smallest eigenvalue $\leq$ harmonic)

see Fig. 7. Using Aouchiche and Hansen's definition of the range, we could not find any counterexample; however, if this initial definition were used, Graffiti 322 would be (too) easily refuted.

The second smallest eigenvalue is $\lambda_{n-1} \approx -1.9659$, and the range of eigenvalues of the gravity matrix is approximately 1.7035.

Cycles of size 21 and 25 also refute Graffiti 197. We conjecture that any cycle of size $1 + 4 * n$ greater than or equal to 17 refutes this conjecture.

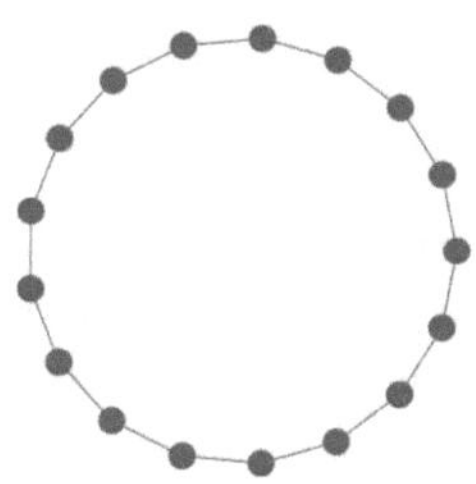

Fig. 7. A counter-example of Graffiti 197 of size 17

5.2 Erratum

We provide a copy of "Written on the Wall" on this project's GitHub, which is otherwise hard to find:

- Graffiti 290 (open) was solved instantly using the definition of gravity in Aouchiche and Hansen's survey [1] (which prompted us to verify the definition). But it was seemingly impossible using the definition from Brewster et al. [2].
  Since the definition of the gravity of a graph is not widespread, here is the correct definition of the gravity matrix from "Written on the Wall" (page 52) and Brewster et al. [2]:
  Gravity Matrix: The matrix (indexed by vertices of the graph) whose $(u, v)th$ entry is 0 if $u = v$ or if there is no path joining u to v; otherwise is:

$$Gr(G) = (1/(n-1))(d(u) * d(v))/d(u, v) \tag{1}$$

  where $d(u)$ is the degree of vertex u and $d(u, v)$ is the distance from vertex u to vertex v in the graph G of size n.
- In the same manner, with Aouchiche and Hansen's definition of the harmonic, Graffiti 140 was refuted by the graph with one edge connecting two vertices. The correct definition is featured in "Written on the Wall" (page 28) and in Favaron et al. [6]:

  Harmonic: The harmonic of a graph is defined as:

$$Hc(G) = \sum_{uv \in E} \frac{2}{d(u) + d(v)} \tag{2}$$

  where and E is the set of edges of a graph G.

- Using Aouchiche and Hansen survey's [1] definition of range, it appeared that our algorithms managed to refute Graffiti 322 (open) with a cycle graph of size 4. Here is the definition of Graffiti 322 from "written on the wall":

  Graffiti 322. If G is a triangle-free graph then the Inverse Even $\leq$ range of eigenvalues of Distance.

  Inverse Even is defined as the sum over the vertices of the inverse of the number of vertices at an even distance from that vertex, i.e.:
  $\sum_{v \in V} 1/Ev(v)$ where $Ev(v)$ is the number of vertices at even distance from vertex v.
  Thus, the cycle of size 4 has 3 distinct distance eigenvalues and an Inverse Even of 4. An error with the definitions seems more likely than this conjecture being left open after dozens of articles with such a simple counterexample.
  The range is defined as the number of distinct values in a vector in Aouchiche and Hansen's survey and (likely inherited from) Favaron et al.'s work, with

no further sources. The range of a vector is usually the difference between the minimum and maximum values. Thus, we think this definition is erroneous but we were not able to find the actual definition of the range used in "written on the wall". The results featured in Table 1 for Graffiti 322 use the usual definition of range; either way, Graffiti 197 is refuted or Graffiti 322 is refuted.
- Graffiti 294 from Aouchiche and Hansen's survey is Graffiti 295 from "written on the wall", we go by this identifier in table 1.
- Graffiti 140 is said to be refuted by Favaron et al. in "residue of a graph" [7] according to "written on the wall", but there is no mention of "harmonic", "deviation", or Graffiti 140 in that paper.

The general lack of sourcing, the loss of multiple articles, and the misleading or erroneous definitions are detrimental to future work on Graffiti conjectures. For instance, we suspect that Graffiti conjectures 28 and 209 may not be refuted as sources are nonexistent (28) or seemingly lost (209). We decided to include Graffiti 140 in the table, unlike 28 and 209, because we could find its refutation source, but we could not find its appearance in the source. Auto-GraphiX conjectures may be more favorable to continuing this work.

6 Conclusion

Search algorithms proved to be powerful ways of refuting conjectures from spectral graph theory, much faster than Wagner's deep cross-entropy method [16]. We can identify three algorithms that seem more effective on the conjectures tested: GBFS, which is overall faster but was not able to refute conjectures in [12] and struggled on Graffiti 30. NRPA, which is very efficient when a repeating pattern refutes a conjecture. And LNMCS/NMCS, which managed to be slightly faster than NRPA on Graffiti 29 and 301, while being faster than the other MCTS algorithms. LNMCS did not outperform NMCS due to the nature of the search space: there rarely is a move that dooms all its child states. It seems that not one single algorithm is able to solve all of the sample of conjectures we selected, and using different approaches is required to adapt to the variety of problems in spectral graph theory conjecture refutation. We recommend trying GBFS, the NMCS, and then NRPA.

Trying to build trees even when the conjecture is applied to any graph can also be helpful, as it reduces the amount of possible builds greatly and focuses on a sub-type of graphs that often features extreme properties over the spectral invariants. It is inexpensive and should be tried first.

However, these methods present limits. Computing score functions that require eigenvalues on big trees (over 300 nodes) can be very costly. They are also dependent on the shape of the score function: a noisy score function with many local minima can be challenging, as well as a score function with more discrete results can lead to an absence of differentiation in the paths to explore. Conjectures requiring computing an NP-hard problem can also severely increase the computing time even for small graphs (30 vertices).

Despite these limitations, and depending on the definition of the gravity matrix, either our program was able to refute up to the state of the art (manually from graph theorists) the conjectures from Graffiti, or it refuted Graffiti 197, a previously open conjecture.

You can access the implementations of the conjecture and the Rust code used to refute them here:

https://github.com/RoucairolMilo/refutationGBR.

Statements and Declarations. Thanks to Édouard Bonnet for fruitful discussions. Thanks to Craig E. Larson and Ermelinda DeLaViña for answering some of our questions and providing "written on the wall" to us.

Competing Interest: The authors have no relevant financial or non-financial interests to disclose

Funding: This work was supported in part by the French government under the management of Agence Nationale de la Recherche as part of the "Investissements d'avenir" program, reference ANR19-P3IA-0001 (PRAIRIE 3IA Institute).

References

1. Aouchiche, M., Hansen, P.: A survey of automated conjectures in spectral graph theory. Linear Algebra Appl. **432**(9), 2293–2322 (2010). https://doi.org/10.1016/j.laa.2009.06.015, https://www.sciencedirect.com/science/article/pii/S0024379509003061
2. Brewster, T.L., Dinneen, M.J., Faber, V.: A computational attack on the conjectures of graffiti: new counterexamples and proofs. Discret. Math. **147**(1–3), 35–55 (1995)
3. Cazenave, T.: Nested monte-carlo search. In: Boutilier, C. (ed.) IJCAI, pp. 456–461 (2009)
4. Cazenave, T.: Generalized rapid action value estimation. In: 24th International Conference on Artificial Intelligence, pp. 754–760 (2015)
5. Delavina, E.: Some History of the Development of Graffiti. In: DIMACS Series in Discrete Mathematics and Theoretical Computer Science 69: Graphs and Discovery, pp. 81–118 (2005)
6. Favaron, O., Mahéo, M., Saclé, J.F.: Some eigenvalue properties in graphs (conjectures of graffiti — ii). Discret. Math. **111**(1), 197–220 (1993). https://doi.org/10.1016/0012-365X(93)90156-N, https://www.sciencedirect.com/science/article/pii/0012365X9390156N
7. Favaron, O., Mahéo, M., Saclé, J.F.: On the residue of a graph. J. Graph Theory **15**(1), 39–64 (1991)
8. Gelly, S., Silver, D.: Monte-carlo tree search and rapid action value estimation in computer go. Artif. Intell. **175**(11), 1856–1875 (2011)
9. Hansen, P., Caporossi, G.: AutoGraphiX: an automated system for finding conjectures in graph theory. Electr. Notes Dis. Math. **5**, 158–161 (2000). https://doi.org/10.1016/S1571-0653(05)80151-9, https://www.sciencedirect.com/science/article/pii/S1571065305801519
10. Kocsis, L., Szepesvári, C.: Bandit based monte-carlo planning. In: Fürnkranz, J., Scheffer, T., Spiliopoulou, M. (eds.) ECML 2006. LNCS (LNAI), vol. 4212, pp. 282–293. Springer, Heidelberg (2006). https://doi.org/10.1007/11871842_29

11. Rosin, C.D.: Nested rollout policy adaptation for monte carlo tree search. In: IJCAI, pp. 649–654 (2011)
12. Roucairol, M., Cazenave, T.: Refutation of spectral graph theory conjectures with monte carlo search. In: International Computing and Combinatorics Conference, pp. 162–176. Springer (2022). https://doi.org/10.1007/978-3-031-22105-7_15
13. Roucairol, M., Cazenave, T.: Solving the hp model with nested monte carlo search, arXiv preprint (2023). arXiv:2301.09533
14. Silver, D., et al.: Mastering the game of go with deep neural networks and tree search. Nature **529**, 484–489 (2016)
15. Silver, D., et al.: Mastering chess and shogi by self-play with a general reinforcement learning algorithm. CoRR abs/ arXiv: 1712.01815 (2017)
16. Wagner, A.Z.: Constructions in combinatorics via neural networks. arXiv:2104.14516 [cs, math] (Apr 2021)

Agent-Based Modeling and Framework Analysis

Framework Showdown: A Practical Benchmarking Study of Agent-Based Modeling Platforms

Antoni Zięciak(✉), Ewelina Badeja, Joanna Wójcicka, Michał Szczygieł, Przemysław Orlikowski, and Jarosław Wąs

Department of Applied Computer Science, Faculty of Electrical Engineering, Automatics, Computer Science and Biomedical Engineering, AGH University of Krakow, al. Mickiewicza 30, 30-059 Krakow, Poland
{zieciak,ebadeja,joannawoj,przorlik}@student.agh.edu.pl, jaroslaw.was@agh.edu.pl

Abstract. In recent years, a wide range of agent-based modeling (ABM) frameworks has been developed. Many of these advanced tools support the creation of complex simulations and offer distinct features for various use cases. This practical comparative study evaluates the capabilities, execution time performance, and operating memory usage of several ABM frameworks through a series of benchmark simulations. The analysis centers on five agent-based modeling (ABM) tools, each correlated with a distinct and widely-used programming language: krABMaga, selected for its computational efficiency; Mesa, designed for its simplicity; MASON, recognized for its robustness; Agents.jl, noted for its ongoing maintenance; and Agent Harmony, which seeks to combine simplicity with high performance. The goal of this work is to provide a clear overview of the available options and assist researchers and practitioners in selecting the best-suited framework for their specific modeling needs.

Keywords: Agent-based modeling · Agent-based modeling frameworks · Software agent · Multi-agent computing

1 Introduction

Agent-Based models (ABMs) are probably the most general approach to the modeling and simulation of complex systems [11]. Such models consist of discrete entities that may interact with each other and the environment. These agents often have predefined rules and the ability to learn and adapt [11]. They can be used to implement systems based on other popular approaches, such as Cellular Automata (CA) and networks. Due to their universality, Agent-Based frameworks have found applications in various subjects, such as the social sciences [4], construction [8], and ecological economics [6]. The popularity and rapid growth of this subject have resulted in the appearance of many Agent-Based modeling

© The Author(s), under exclusive license to Springer Nature Switzerland AG 2026
K. Mason et al. (Eds.): IASE 2025, CCIS 2976, pp. 31–47, 2026.
https://doi.org/10.1007/978-3-032-28263-7_3

frameworks [10]. Comparisons between different frameworks have been made in order to answer the question of which one to choose [5,12].

This work focuses on the comparison of the functionalities crucial for the development of Agent-Based models. As such, we count: representation of the simulation state, agent creation, scheduling, and representation of a simulation space. We have also checked the possibilities of visualization and unique features of each framework. The compared solutions are: MASON [9], krABMaga [1], Mesa [7], Agents.jl [3] and Agent Harmony (AgH) [13]. In addition to the features offered, the aforementioned solutions were also tested in terms of performance and memory efficiency.

2 Frameworks Comparison

Five frameworks covered in this work – MASON, krabMaga, Mesa, Agent Harmony, and Agents.jl – were chosen because of their popularity and the fact that each of them is implemented in a different programming language. Additionally, each framework has unique features and qualities that could be useful in an agent-based approach.

2.1 Overview

MASON is a high-performance agent-based simulation library core in Java. It is designed to serve as the foundation for large custom-purpose Java simulations and also to provide extensive functionality for a wide range of lightweight simulation needs. MASON contains both a model library and an optional suite of visualization tools in 2D and 3D. It is the most established of the comparable frameworks and includes thorough documentation.

krABMaga is a discrete event simulation engine for developing ABM simulations written in the Rust programming language. It is designed to be a ready-to-use tool for the ABM community, and for this reason, the architectural concepts of the well-adopted MASON library were re-engineered to take advantage of the Rust peculiarities and programming model. Notably, the visualization and the simulation subsystems remain fully separate.

Mesa is an Apache2-licensed ABM Python framework. It enables users to easily create agent-based models with built-in core components or customized implementations, visualize them with a browser-based interface, and analyze the results using Python's data analysis tools. Mesa's objective is to facilitate the comprehension and resolution of complex problems by democratizing access to simulations for a broad audience. The code under review exhibits the hallmarks of Python's clean code structure; however, it is deficient in the area of proper documentation.

Agent Harmony already demonstrates impressive computational performance and relatively low memory usage. It is implemented in C++.

The final framework under consideration is Agent.jl, a pure Julia framework for ABM. The software in question boasts a number of notable advantages, including its rapid execution, its minimal learning curve, and its requirement for just a small amount of code. The software is accompanied by extensive documentation, which includes tutorials, illustrative examples of ABM implementations, and integration examples with other Julia packages.

2.2 Representation of Simulation State

Across ABM frameworks, the simulation state is typically encapsulated in a central structure, commonly referred to as Model, State, or SimState, which serves as the single source of truth. It encompasses all the essential data needed to describe and track the evolving behavior of the system. This usually includes the agents involved in the simulation, the space in which they move and interact, and various other parameters that influence the dynamics of the simulation.

Agent Harmony proposes a more flexible approach by defining a base responsible for agent storage, thereby allowing users to define additional components needed for their specific simulations, such as a simulation space or custom properties. Furthermore, this framework employs a distinct approach to the simulation state, deviating from a centralized execution point. Conversely, the Model class is passed as an argument to each agent's step method and also referenced within the Schedule class. This approach enables a more distributed management of the simulation's data and logic.

MASON employs a comparable methodology. The simulation state is stored in a single instance of the *SimState* class. This object is responsible for the management of the global simulation schedule, the operation of random number generators, and the storage of agent states and environmental variables in both built-in and user-defined data structures. In each iteration, the *SimState* object reference is passed to each agent's *step* method. This approach confers a substantial degree of autonomy in delineating the parameters of the simulation environment.

MESA, Agents.jl and krABMaga employ a step method in simulation state for handling updates at each step. However, krABMaga exposes this method as additional logic that executes once for each simulation step. Moreover, krABMaga is equipped with supplementary functionality that facilitates the implementation of customized methodologies. These methodologies permit users to introduce specific logic prior to and following each simulation step. Additionally, users can define a condition that determines the simulation's termination.

2.3 Agents Creation and Representation

The manner in which agents are generated within an agent-based modeling framework exerts a fundamental influence on its design and capabilities. Flexible agent definitions enable users to customize behavior and structure, thereby supporting a wide range of modeling scenarios. In contrast, stricter agent structures promote consistency and tighter integration with the framework's components. These design choices directly influence how models are built, extended, and executed.

The selected frameworks utilize certain common approaches to agent creation and representation, in order to align with the fundamental principles of agent-based modeling. Agents are identified by a unique identifier, and, when relevant, they are positioned in space. Additionally, they possess custom attributes that capture their internal state. A standardized step function governs the agent behavior at each simulation tick, and integration with the scheduler ensures consistent execution.

The majority of frameworks introduce an Agent abstraction through traits, interfaces, or base classes. This standardizes agent behavior and facilitates integration with the simulation engine. The Agent Harmony framework constitutes an exception, as it provides a more flexible approach to the definition of agents. It enables the user to designate an *agent*(scheduled object) as any structure that possesses mandatory methods.

Furthermore, MASON implements an additional class *Steppable*, which broadens the capabilities of an Agent's behavior and user control. It facilitates the establishment of lifecycle hooks, empowering users to specify customized behavior at crucial junctures within the simulation process. KrABMaga also supports lifecycle hooks, although in a more constrained form. In KrABMaga, these hooks are confined to actions executed immediately prior to or subsequent to an agent's step function, thereby offering a more constrained scope for customization.

In the Mesa framework, an agent's class must be derived from the specified class *Agent* and is required to implement the function *step*. The constructor of the base class autonomously assigns an identifier to the created agent and integrates it into the designated simulation model. Conversely, in Agents.jl, an agent in its most basic form only necessitates an identifier, with its behavior being defined by an external function.

2.4 Scheduling

Scheduler is a mechanism that controls the order and manner in which agents are activated during each time step of the simulation. In MASON, krabMaga, and Agent Harmony, it is based on a priority queue. MASON facilitates the scheduling of agent activation with an interval. It also enables the initiation of Agent Harmony calls and optionally advances (aiding in the management of synchronous calls). However, these are the only significant distinctions between the two systems.

In Agent.jl, the scheduling process is managed by pure functions that return an iterator of agent IDs to be activated at each simulation step. A scheduler is a function: The current model is taken as the input, and the list or iterator of agent IDs to activate is returned. The allocation of schedulers can be executed during the model's creation phase or can be supplied in a dynamic manner at runtime. The framework incorporates several predefined schedulers, including random_activation, fastest, and by_id. However, it also permits the implementation of custom schedulers through the use of plain Julia code.

Mesa does not contain a Schedule-like class. The class designated *Model* contains a method *step* that is invoked in each simulation step. Any non-standard behaviors must be implemented within the agent's class.

2.5 Representation of Space

Another important part of the ABM framework implementation is space. The creators of the framework have addressed the significant demand for a range of space representations, offering a multitude of options to choose from for the best possible models:

- **Grid (Discrete)**: The space is divided into square cells, and agents occupy and move between discrete positions on a 2D grid.
- **Continuous Space**: Agents move freely using real-valued coordinates, which allows for smooth motion.
- **Network Space**: Agents are placed on nodes of a graph and can interact or move along edges based on network topology.
- **Nested Space**: A hierarchical structure in which one space contains multiple subspaces or layers, enabling complex, multi-level environments.
- **Hexagonal Grid**: A grid composed of hexagonal cells instead of squares, providing more uniform neighbor distances and directions.
- **Toroidal (wraparound)**: The spatial configuration exhibits a wrapping effect along its periphery, resulting in agents exiting from one side re-entering from the opposite side, akin to the structural characteristics of a torus.
- **Multi-dimensional**: Extends beyond 2D, enabling agents to exist and interact in 3D or higher-dimensional spaces.

The following table provides a comprehensive overview of the types of simulation space supported by each covered framework Table 1.

Table 1. Comparison of Space Types in Different Frameworks

Space Type	KrabMaga	Mesa	Mason	Agent Harmony	Agents.jl
Grid (Discrete)	Yes	Yes	Yes	Yes	Yes
Continuous Space	Yes	Yes	Yes	Yes	Yes
Network Space	Yes	Yes	Yes	Yes	Yes
Nested Space	No	Limited	Yes	Yes	No
Hexagonal Grid	No	No	Yes	No	Yes
Toroidal (Wraparound)	Yes	Yes	Yes	Partially	Yes
Multi-dimensional	No	No	Yes	No	Yes

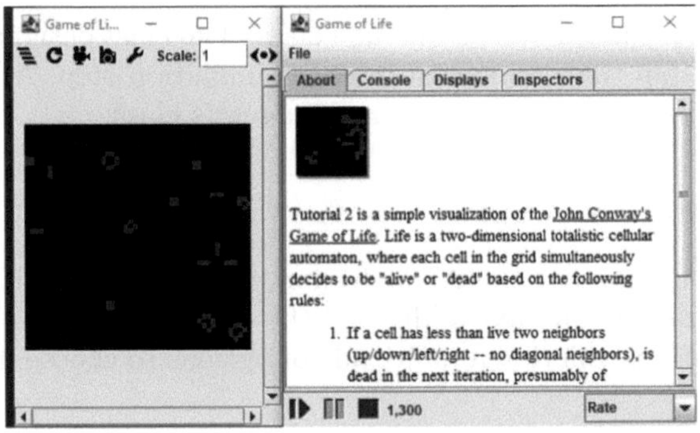

Fig. 1. Example of 2D visualization in MASON

2.6 Visualization

A comparison between the discussed frameworks reveals significant disparities in the utilization of visualization techniques. The MASON framework is predicated on the principle of maintaining a distinct separation between the simulation model and its visual representation. This framework facilitates the dynamic attachment or detachment of models from visualizers, thereby enabling platform modifications during the execution of the program. Visualization can be performed in two dimensions 2D (Fig. 1) or three dimensions 3D (Fig. 2), with the latter requiring the Java3D toolkit. In both cases, the software facilitates interaction with the model, adjustment of parameters, and the recording of alterations. While the potential applications are extensive, MASON's extensive array of visualization object types might present challenges for individuals unfamiliar with such software.

Mesa's visualization module employs Solara for rendering in Jupyter notebooks or as standalone web applications. The system under consideration has the capacity to support a variety of visualizations, including matplotlib plots, agent grids, and custom visualization components (Fig. 3).

Visualization in Agents.jl is built around Plots.jl. It provides flexible and customizable 2D plotting for agent-based models based on space types DiscreteSpace, ContinuousSpace, and GraphSpace (Fig. 4). Users can employ any Plots.jl-compatible backend, such as GR, PyPlot, or Plotly, offering a variety of rendering styles and output formats including interactive plots, static images, and PDF exports. For fully interactive model exploration, Agents.jl integrates with the InteractiveDynamics.jl package and Makie.jl plotting backend. Agents.jl supports both static plots and dynamic, interactive dashboards within a single workflow.

In krabMaga, visualization can be done in two ways. The first option uses the Bevy engine for a native visualization and monitoring system (Fig. 5), while the second leverages WebAssembly to display the simulation directly in a web browser. KrabMaga's unique feature in this field is to visualize an agent on a square grid, providing the agent's color is not enough – it needs to have a square

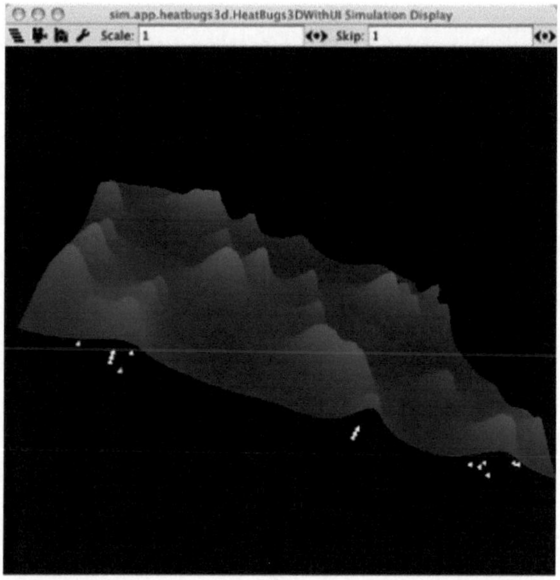

Fig. 2. Example of 3D visualization in MASON. Source: https://cs.gmu.edu/~eclab/projects/mason/docs/tutorial0/

emoji assigned, which is stored in the assets/emojis folder. Keep in mind that it hasn't been officially released yet – it's currently a release candidate.

In the event that visualization issues arise within the aforementioned frameworks, it is recommended to refer to the available tutorials or official implementations of various models, which can be found on their official webpages.

The visualization module of Agent Harmony is still in its nascent stages of development, offering a limited selection of options at this point. Nevertheless, the process of visualizing a simple grid is uncomplicated and is the most accessible among the covered frameworks (Fig. 6).

2.7 Additional Unique Features

The previously described functionalities should suffice for implementing every ABM simulation, however, certain frameworks have opted to augment their features, thereby enhancing the efficacy of computing processes in terms of both speed and resilience. In order to achieve enhanced performance, Mesa, MASON, Agents.jl have implemented concurrent processing. Mesa, Agents.jl and MASON facilitate concurrent processing of batches of simulations; however, it should be noted that Mesa's concurrent processing feature is exclusively available in native Python, not in the Jupyter Notebook.

Furthermore, MASON, through its ParallelSequence functionality, enables the execution of numerous parallel steps within a single simulation. A similar outcome can be accomplished in Agents.jl. In the absence of shared memory

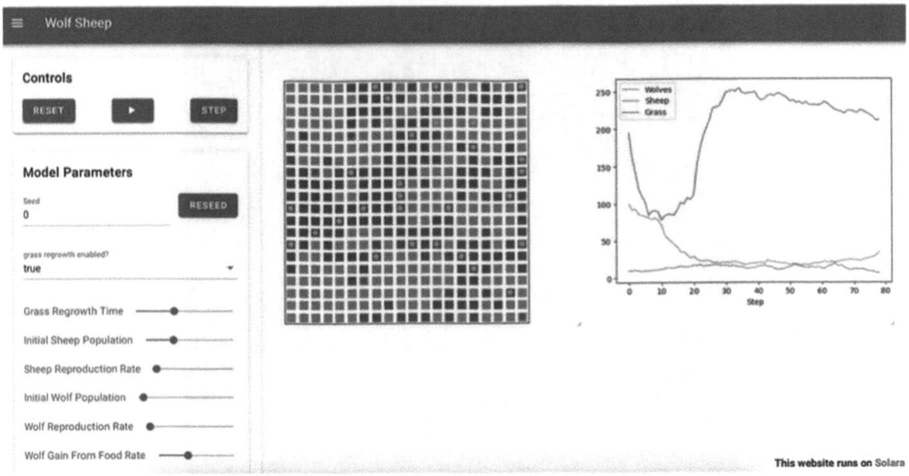

Fig. 3. Example of visualization in Mesa. Source: https://mesa.readthedocs.io/latest/

conflicts, the process of stepping logic can be manually parallelized through the utilization of Julia's concurrency tools.

With regard to krabMaga, parallel computing remains an experimental feature.

Another mechanism that has been demonstrated to accelerate the execution of simulations is distributed computing. The distribution and execution of tasks in parallel is a capability that can be implemented across numerous systems, whether on a single machine or across multiple ones. The following text is intended to provide a comprehensive overview of the subject matter.

Distributed MASON is in the early stages of testing. Currently, it only supports 2D bounded or toroidal grids, GIS/GeoMason, and continuous space. Sparse grids, 3D fields, networks, unbounded fields, and multiple fields of different sizes are not yet implemented. Consequently, significant progress is required before it can be considered safe for practical use. Agents.jl, by contrast, leverages Julia's Distributed module to achieve concurrency, enabling its parallel simulations to transfer readily to distributed systems.

The final unique feature is MASONS's integrated checkpoint system. This feature facilitates the expeditious and straightforward preservation of the current simulation's state, a crucial aspect that is particularly advantageous in the event of a system crash. Later saved checkpoints can be loaded or used for data analysis. A similar system can be manually implemented in all the other covered frameworks; however, it may require a significant investment of effort for complicated environments.

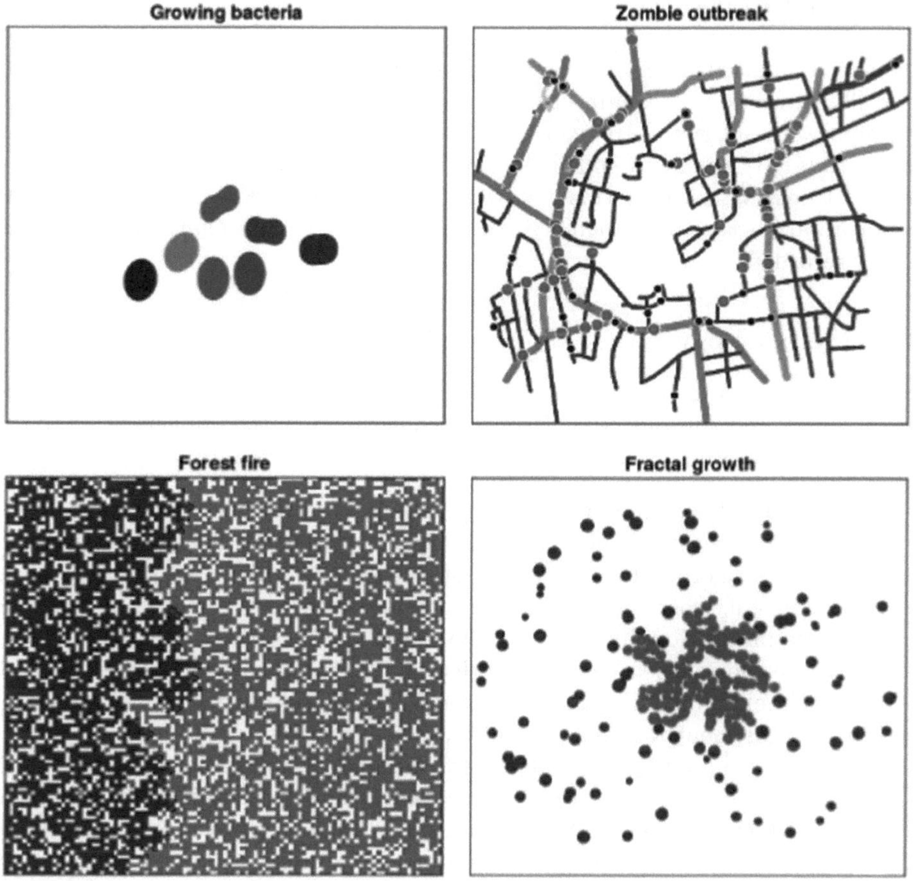

Fig. 4. Examples of visualization in Agents.jl. Source: https://juliadynamics.github.io/Agents.jl/stable/

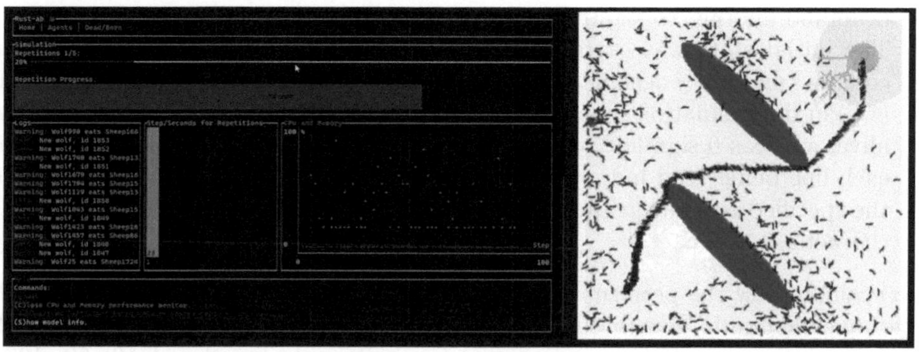

Fig. 5. Example of visualization in krabMaga with simulation monitor. Source: https://krabmaga.github.io/

3 Benchmarks

We have evaluated the frameworks'performance on a specific Conway's *Game of Life* [2] simulation. A comprehensive evaluation was conducted in which both variants were subjected to rigorous testing, incorporating both graphical user interface (GUI) scenarios and those devoid of a GUI. A series of tests were conducted to assess the performance of the system under various conditions. The mean iteration time and memory consumption for each scenario were meticulously recorded.

3.1 Methodology

In each of the aforementioned frameworks, Conway's *Game of Life* [2] was implemented with and without GUI. To ensure consistency in the implementation, the following assumptions were made:

- Every grid cell is depicted as a distinct agent, resulting in a one-to-one correspondence between the number of agents and grid cells. This alignment facilitates more rigorous benchmarking processes, particularly regarding aspects such as agent scheduling, where the number of agents required to schedule increases proportionally with the size of the grid.
- The simulation space is designed with a periodic boundary condition to ensure that all agents, including those located at the map's edge, consistently maintain an equivalent number of neighboring cells.
- Synchronous updates are employed to guarantee a uniform grid configuration subsequent to each iteration. Initially, all agents are required to determine their next state by considering both their current state and the states of neighboring agents. Once this computation is concluded for all agents, they are obligated to transition to the newly determined state.
- The graphical representation of the simulation is as minimalist as possible – simple grid populated with the appropriate number of cells. Each cell within this grid is assigned one of two distinct colors: one to indicate that the cell is dead and another to signify that the cell is alive.
- The initial state was extracted from a text file where the character in the column i and row j designated the initial state of an agent localized at position (i, j) in the simulation space. A character value of 1 indicated that the cell was alive, whereas 0 signified otherwise. The file was processed sequentially, with each line being read individually. During parsing, an agent corresponding to the specified cell was instantiated, reflecting its initial state as denoted by the character.

To evaluate how the performance of the frameworks scales with changes in grid size, five randomly generated maps of varying sizes were created. The initial states were constructed as square matrices with side lengths of 10, 50, 100, 250, and 500. Each matrix was subjected to three distinct iteration counts to assess the impact of initial state loading on the overall simulation performance.

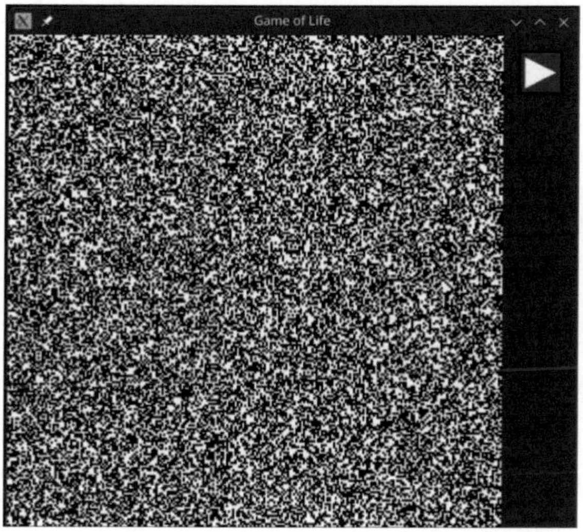

Fig. 6. Example of visualization in Agents Harmony

Consequently, the grid of size 10×10 was executed for 1000, 10000, and 100000 iterations. Grid size 50×50 was tested for 100, 1000, and 10000 iterations, while all other grids were evaluated for 10, 100, and 1000 iterations.

The Python script was developed to facilitate the automated execution of simulations. For the implementation without visualization, each scenario was executed sequentially multiple times to ensure that the experiment was conducted under optimal conditions. For implementations incorporating a graphical interface, manual intervention was necessary due to the presence of control buttons within the frameworks, which are tasked with initiating the simulations.

3.2 Speed Performance

In the implementation without a graphical user interface, the time measurement encompassed the entire sequence of executions for a specified grid size across a predetermined number of iterations. The mean time per iteration was then derived by dividing the total measured time by the number of iterations and subsequently by the count of simulation re-runs. This methodology introduces overhead because it considers the time required to execute sub-processes and loading the initial state is considered as part of the iteration execution time. However, as illustrated in Fig. 7a, the impact of this overhead diminishes with larger map sizes and an increased number of iterations.

As illustrated in Fig. 7a, Agent Harmony exhibits the highest performance in scenarios without visualization, surpassing all other frameworks in each scenario. The only framework whose results are comparable—though not superior—is krABMaga, which demonstrates slower performance by 31%, 43%, 72%, 83%, and 198% in progressively larger spaces. Notably, krABMaga's mean iteration

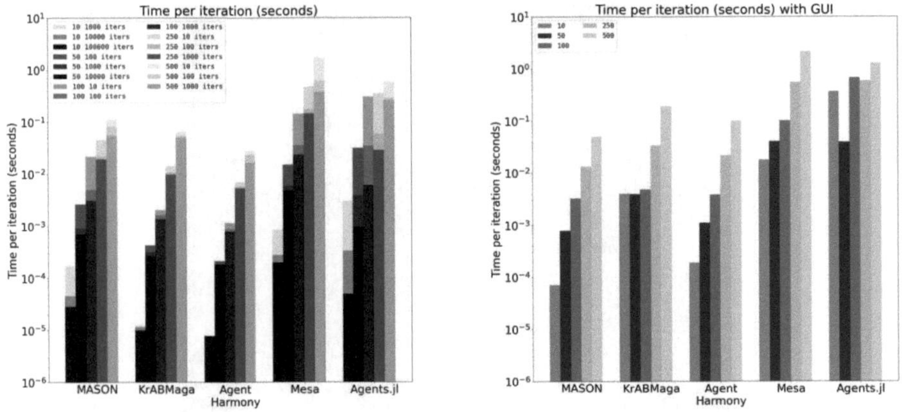

(a) Mean iteration time in each scenario, without visualization

(b) Mean iteration time, averaged between iteration counts, with visualization

Fig. 7. Mean iteration time for each framework

time for a grid of size 500 × 500 approximates that of MASON's (krABMaga: 0.050 s for 1000 iterations, MASON: 0.055 s) more closely than it does Agent Harmony's (0.017 s for 1000 iterations). Mesa exhibits the poorest performance as iteration numbers increase. However, for grids of sizes 10 × 10, 50 × 50, and 100 × 100 with the smallest number of iterations, Agents.jl displays a higher mean iteration time. This is likely attributable to the initial time-consuming process required for Julia to load the necessary environment to execute the simulation.

In the context of timing evaluation for the graphical interface versions, the automated timing mechanism employed via the Python script was deemed unsuitable because the graphical user interface necessitated manual initiation of the simulation process. Consequently, the timing assessment was integrated directly into the implementation. This timing captured solely the simulation execution phase, excluding the loading phase of the initial state.

Figure 7b presents the mean execution time of an iteration, averaged across multiple iterations, under the condition that data loading is excluded from the measurements. A comparison with the implementation devoid of a graphical user interface reveals a change in ranking. MASON now occupies the premier position, surpassing AgH by factors of 63%, 29%, 14%, 37%, and 51% across progressively larger spaces. The KrABMaga framework exhibits a fixed maximum iteration time, yielding comparable results for dimensions 10 and 50. However, variations are observed for larger spaces, where KrABMaga attains performance levels similar to those of Agent Harmony. The lowest performance, contingent upon the grid size, is attributed to Mesa and Agent.jl. Notably, the latter framework is the unique one that demonstrates a non-monotonic growth in mean iteration time.

Figure 8 shows an analysis of mean iteration times with and without a graphical user interface (GUI), expressed as a percentage. As expected, MASON demonstrates superior performance. Interestingly, certain scenarios reveal that solutions with a GUI have a lower mean iteration time compared to those without a GUI, which is likely attributed to the exclusion of data loading measurements in the timings for the GUI. Conversely, Agents.jl exhibits the most significant overhead upon the integration of a graphical interface, amounting to $10^6\%$.

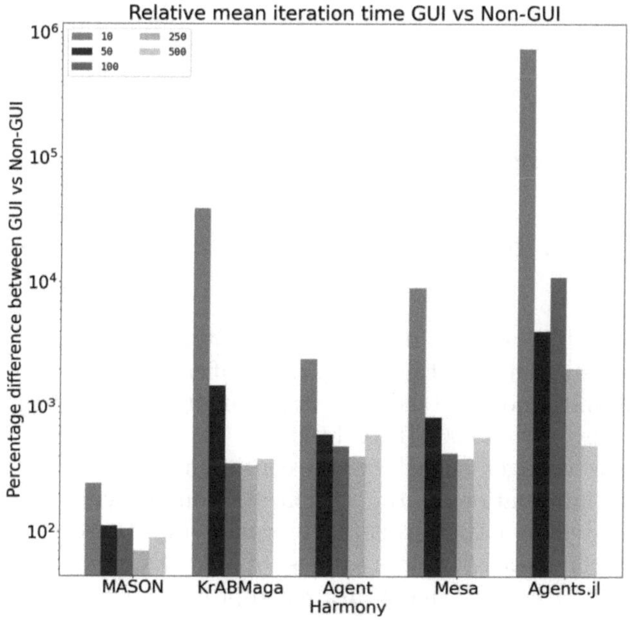

Fig. 8. Relative mean iteration time between both versions

3.3 Memory Consumption

Memory consumption data were systematically gathered for both implementations, with and without the inclusion of visualization. Each benchmark scenario was executed by a Python script in a sub-process. While the simulation was active, memory consumption information was collected using the Linux program *pmap* to account for all child processes of a given process.

Figure 9a presents data regarding the average memory usage of each framework, relative to grid size, considering solely the maximum number of iterations, without any visual representation provided. It can be observed that Agent Harmony exhibits optimal performance, closely followed by krABMaga. The disparity in memory consumption between these two frameworks is marginal with smaller grids (18% for a grid size of 10×10), escalating to 221% for the largest

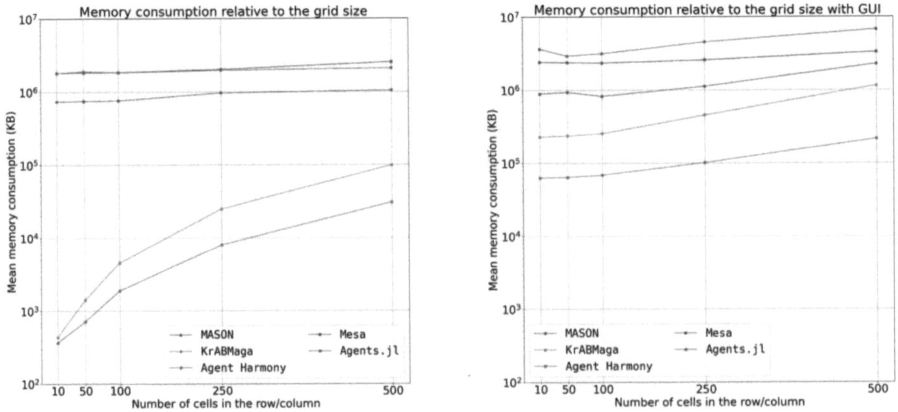

(a) Mean memory consumption without visualization

(b) Mean memory consumption with visualization

Fig. 9. Mean memory consumption for each framework

grid. While other frameworks demonstrate greater stability in memory usage, they are more demanding in this context. Specifically, Agents.jl and Mesa consume a comparable amount of memory, approximately 2 GB, whereas MASON's consumption ranges from 600 MB to 1 GB.

The analysis changes slightly when considering Fig. 9b, which illustrates memory usage information in the visualization-implemented scenario. Agent Harmony consistently outperforms other frameworks. However, the results for krABMaga are comparatively close to those of MASON, notably with larger grid sizes. Specifically, Agent Harmony utilizes approximately 215 MB of RAM for a grid encompassing $250,000$ agents, whereas krABMaga requires slightly over 1 GB, with MASON utilizing 2.2 GB. Moreover, when evaluating the graphical user interface, Agents.jl appears to be more memory-intensive than Mesa, requiring approximately 6.7 GB of RAM for the largest grid configuration.

Among the frameworks evaluated, krABMaga exhibits the highest percentage increase, requiring 520 times more memory for a 10×10 grid size. In terms of memory consumption (Fig. 10) krABMaga is followed closely by Agent Harmony. These two frameworks are unique in that they demonstrate a reduction in memory consumption overhead as the grid size increases. In contrast, both MASON and Agents.jl show relatively constant memory consumption for smaller grids, with a notable increase as grid size enlarges. Notably, Mesa is the only framework exhibiting consistent and relatively low memory overhead following the addition of the GUI.

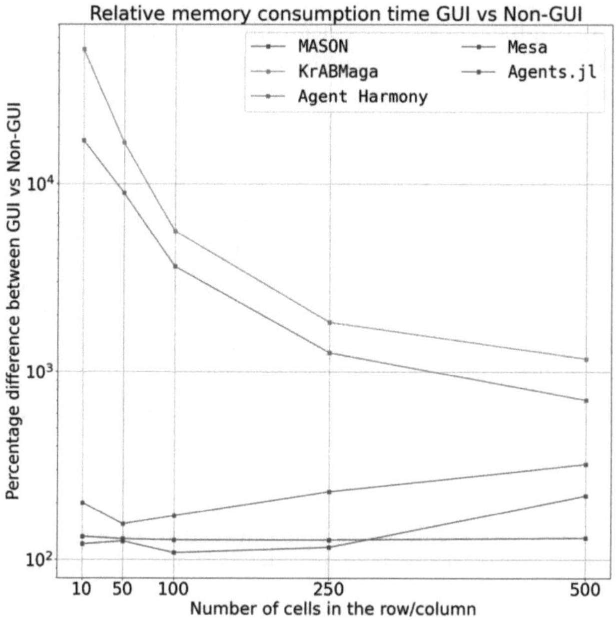

Fig. 10. Relative mean memory consumption between both versions.

4 Conclusion

There are many factors to consider when choosing a framework for ABM implementation, including the preferred programming language, the complexity of the planned model, and performance requirements. Each of the surveyed Agent-Based Modeling (ABM) frameworks is capable and flexible in its own way, but they differ significantly in performance, usability, and extensibility.

- **MASON** framework has reached a state of maturity and has been thoroughly documented, providing a high degree of flexibility in the design of environments. It occupies a median position with regard to memory and performance; however, it is noteworthy for its substantial visualization capabilities. However, its richness can prove overwhelming. The visualization system is characterized by its modular design, which introduces a relatively modest overhead compared to alternative approaches.
- **Mesa** is one of the slower and more memory-intensive frameworks. Notably, it is distinguished by its seamless integration with Python's extensive ecosystem, encompassing support for Jupyter Notebooks and data tools such as `pandas`. However, the documentation is limited in scope and there is an absence of a scheduler that can be customized, which represents a significant constraint.
- **krABMaga** draws inspiration from MASON. This framework has been shown to exhibit superior memory efficiency, ranking second only to Agent Harmony. Additionally, it delivers high performance when utilized in the

absence of a graphical user interface. The visualization module is distinguished by its unique features, which, while not immediately apparent, can be mastered with sufficient effort.
- **Agents.jl** is particularly well-suited framework for experienced Julia users, as it facilitates the extension of the framework using Julia's language features. A notable strength of the platform is its native support for distributed computing, a feature that sets it apart from many other systems on the market. However, it has been observed to exhibit tendencies toward memory exhaustion and diminished processing speed in scenarios involving extensive simulations.
- **Agent Harmony** is still in the early stages of development. However, preliminary findings indicate that it is both extremely fast and lightweight in terms of memory usage. At present, the visualization capabilities are rudimentary; however, this simplicity also renders them straightforward to implement and extend.

Regardless of the choice, readers are encouraged to experiment with different tools and dive deeper into the rich and evolving field of agent-based modeling.

Acknowledgments. The research was financed by a scientific subsidy from the AGH University of Krakow.

Disclosure of Interests. The authors declare that they have no known competing financial interests or personal relationships that could have influenced the work reported in this paper.

References

1. Antelmi, A., et al.: On evaluating rust as a programming language for the future of massive agent-based simulations. Commun. Comput. Inf. Sci. **1094**, 15–28 (2019). https://doi.org/10.1007/978-981-15-1078-6_2, Conference of 19th Asia Simulation Conference, AsiaSim 2019; Conference Date: 30 October 2019 Through 1 November 2019; Conference Code:233729
2. Conway, J., et al.: The game of life. Sci. Am. **223**(4), 4 (1970)
3. Datseris, G., Vahdati, A.R., DuBois, T.C.: Agents.jl: a performant and feature-full agent-based modeling software of minimal code complexity. Simulation, 003754972110688 (2022). https://doi.org/10.1177/00375497211068820
4. El-Sayed, A.M., Scarborough, P., Seemann, L., Galea, S.: Social network analysis and agent-based modeling in social epidemiology. Epidemiol. Perspect. Innov. **9**(1), 1–9 (2012)
5. Gacek, P., et al.: A comparison of selected agent-based modelling frameworks. In: Wyrzykowski, R., Dongarra, J., Deelman, E., Karczewski, K. (eds.) PPAM 2024. LNCS, vol. 15581, pp. 276–288. Springer, Cham (2024). https://doi.org/10.1007/978-3-031-85703-4_19
6. Heckbert, S., Baynes, T., Reeson, A.: Agent-based modeling in ecological economics. Ann. N. Y. Acad. Sci. **1185**(1), 39–53 (2010)

7. Kazil, J., Masad, D., Crooks, A.: Utilizing Python for agent-based modeling: the mesa framework. In: Thomson, R., Bisgin, H., Dancy, C., Hyder, A., Hussain, M. (eds.) SBP-BRiMS 2020. LNCS, vol. 12268, pp. 308–317. Springer, Cham (2020). https://doi.org/10.1007/978-3-030-61255-9_30
8. Khodabandelu, A., Park, J.: Agent-based modeling and simulation in construction. Autom. Constr. **131**, 103882 (2021)
9. Luke, S., et al.: The MASON simulation toolkit: past, present, and future. In: Davidsson, P., Verhagen, H. (eds.) MABS 2018. LNCS (LNAI), vol. 11463, pp. 75–86. Springer, Cham (2019). https://doi.org/10.1007/978-3-030-22270-3_6
10. Pal, C.V., Leon, F., Paprzycki, M., Ganzha, M.: A review of platforms for the development of agent systems. arXiv preprint arXiv:2007.08961 (2020)
11. Sayama, H.: Introduction to the Modeling and Analysis of Complex Systems (2015)
12. Wrona, Z., et al.: Comparison of multi-agent platform usability for industrial-grade applications. Appl. Sci. **14**(22) (2024). https://doi.org/10.3390/app142210124, https://www.mdpi.com/2076-3417/14/22/10124
13. Zięciak, A.: Agent Harmony (2025). https://github.com/Tosiekdev/Agent-Harmony

Comparison of the Algorithms Used in Agent-Based Modeling Frameworks

Antoni Zięciak(✉), Przemysław Orlikowski, and Jarosław Wąs

Department of Applied Computer Science, Faculty of Electrical Engineering, Automatics, Computer Science and Biomedical Engineering, AGH University of Krakow, al. Mickiewicza 30, 30-059 Krakow, Poland
{zieciak,przorlik}@student.agh.edu.pl, jaroslaw.was@agh.edu.pl

Abstract. Agent-Based Modeling (ABM) is a powerful approach for simulating complex systems composed of interacting autonomous agents. A variety of ABMs exist, each offering different design paradigms and performance characteristics. In this study, we analyze and compare the core algorithms employed by five distinct ABM frameworks - KrABMaga, Mesa, MASON, Agents.jl, and Agent Harmony, a framework developed in C++. To ensure fair and consistent evaluation, we re-implemented the core algorithms, functions, and classes from KrABMaga, Mesa, MASON, and Agents.jl in C++ and conducted extensive performance tests. The findings indicate significant trade-offs: Agent Harmony consistently surpasses other models in neighborhood and spatial queries due to its optimized indexing; Mesa exhibits notable versatility, excelling in agent movement within continuous spaces, albeit with delayed performance in removal tasks; MASON demonstrates strong performance in fundamental space operations and scheduling; Agents.jl achieves efficient movement and removal within large grids and excels in network neighbor discovery; whereas KrABMaga, despite its limited neighborhood support, efficiently executes movement and connection operations. The comparison provides valuable insights for researchers and developers seeking optimal frameworks for high-performance, large-scale agent-based simulations.

Keywords: ABM · agents · agent-modelling · KrabMaga · Mesa · comparison · MASON · Agents.jl · Agent Harmony

1 Introduction

Agent-Based Modeling (ABM) has become a method in many disciplines, including the social sciences [5], construction [10], and crowd dynamics [17]. As systems grow in complexity and scale, the computational performance of ABM frameworks becomes increasingly critical – especially when simulations involve thousands of millions of agents. These frameworks facilitate the development of systems grounded in other widely recognized methodologies, such as Cellular Automata and network models. Various frameworks have been compared to determine which one should be selected [7,16]. Despite the abundance of ABM

© The Author(s), under exclusive license to Springer Nature Switzerland AG 2026
K. Mason et al. (Eds.): IASE 2025, CCIS 2976, pp. 48–66, 2026.
https://doi.org/10.1007/978-3-032-28263-7_4

tools, there remains a lack of detailed comparisons of the underlying algorithmic structures, their performance implications, methods, and environments available for conducting simulations. This paper addresses that gap by comparing five ABM frameworks, KrABMaga [1], Mesa [9], Agents.jl [3], MASON [12] and Agent Harmony – with a focus on their core simulation algorithms. To ensure a consistent, reproducible, and meaningful comparison, we re-implemented the core simulation of all frameworks to C++. This eliminated differences due to language-level performance and allowed us to isolate the impact of algorithmic design on execution speed, scalability, and memory usage. Through a series of benchmark tests, we evaluated the performance of each framework under various agents and interaction loads.

The remainder of this paper outlines the algorithmic foundations of each framework, details the C++ implementation, and presents a thorough comparative analysis of their performance characteristics.

2 Framework Comparison

2.1 Overview

This section provides an overview of the five ABM frameworks evaluated in this study. KrABMaga, Mesa, MASON, Agent.jl and Agent Harmony. Each framework reflects a different design philosophy, programming paradigm, and makes different trade-offs between usability and performance.

KrABMaga is a high-performance, discrete-event Agent-Based Modeling (ABM) framework implemented in Rust. Drawing inspiration from the architectural concepts of the popular Agent-Based Modeling framework MASON, KrABMaga reimagines these principles within Rust's ownership-based memory model. Unlike simulations and visualizations, KrABMaga distinctly separates its subsystems, allowing for more efficient runtime behavior and clearer modularity. Designed as a ready-to-use tool for ABM simulations, KrABMaga focuses on speed, and flexible agent behavior design.

On the contrary, Mesa, is a Python-based ABM framework that prioritizes accessibility and extensibility. Its modular architecture and built-in browser interface allow for rapid model development and visualization. Mesa is particularly well-suited for educational purposes, prototyping, and exploratory modeling. However, it is not optimized for large-scale or performance-critical solutions.

Agent Harmony (AgH) [18] is a lightweight ABM framework developed in C++ as part of this research. Although it has not been officially released, Agent Harmony is designed to push the boundaries of computational efficiency and scalability in agent-based simulations. Agent Harmony emphasizes low overhead, direct memory control, and algorithmic extensibility, making it particularly well-suited for high-performance scenarios. The framework aims to bridge the gap between usability and execution speed by offering a minimalist yet powerful architecture for modeling complex agent interactions.

MASON is a mature, high-performance agent-based simulation toolkit. It is written in Java. It was developed to support large-scale simulations and provide a rich model and visualization ecosystem. MASON includes both a flexible core simulation engine and optional 2D/3D visualization modules that are especially useful for debugging and presentations. MASON's extensive documentation and long-standing community make it one of the most reliable ABM frameworks available. Its object-oriented design allows for clean simulation logic and reusability.

Agents.jl is a modern, pure Julia framework for agent-based modeling that emphasizes simplicity and speed of execution. Its streamlined syntax and advanced meta-programming features facilitate the succinct model formulation with minimal boilerplate code. Agents.jl benefits from Julia's high-performance execution and seamless integration with the broader Julia data science ecosystem. The framework is well-documented and includes numerous tutorials and examples, covering a range of domains from social science to epidemiology. Its ease of use and runtime efficiency make it an excellent choice for both researchers and practitioners seeking a fast-expressive ABM tool with minimal overhead.

Together, these frameworks represent three distinct points on the spectrum of ABM tool design: from the performance-oriented (KrABMaga, Agent Harmony), through the user-friendly and flexible (Mesa), the established robustness of Java (MASON), to the high-level speed and clarity of Julia (Agents.jl). This comparison establishes the foundation for a comprehensive analysis of their algorithmic designs and run-time behavior.

2.2 Simulation State

SimState model is a simulation model within MASON. The model is equipped with the functionality to pause the simulation and subsequently resume it from a designated checkpoint. The core logic responsible for executing the simulation is encapsulated in the doLoop method. The class designated as *SimState* is equipped with a schedule that is responsible for ensuring the execution of agent updates in the correct sequence.

In the Mesa framework, the simulation state is stored in the derived class of the *Model* class. Agents are maintained in a private field as an instance of the *AgentSet* class. They are also stored in the dictionary mapping each agent type to the appropriate agent set. The *step* method constitutes the principal mechanism through which the model executes its processes.

Within the KrABMaga framework, the principal modeling class is the *SimState* class, accompanied by a distinct scheduler. The *SimState* class incorporates *before_*, and *after_step* methods. However, it is within the *Schedule* class that the actual simulation execution resides, manifested in the *step* method. The *priority* mechanism is employed to store and manage agents' execution.

In Agent.jl, the Abstract type *StandardABM* is employed as the most popular class representing the simulation state. The following parameters are used to define the model:

- type of the simulation space,
- the common type among agents in the model,

- the type of the container storing agents – it's either a dictionary or a vector,
- the type of a function responsible for updating the model,
- the type of a function responsible for stepping an agent,
- the type of a function responsible for agents' scheduling.

The Agents are stored in a specialized container class. The 'step!' method assumes responsibility for conducting the simulation, supported by a *scheduler* function.

In the Agent Harmony framework, the class responsible for maintaining the simulation state should inherit from struct *Model*. This facilitates the segregation of agents into distinct containers based on their types. The class is parameterized according to the specific agent types it encompasses. Each agent type is required to implement a function that indicates whether the agent is active. Oversight and execution of the simulation are entrusted to the class denoted as *Schedule*.

2.3 Agent Methods

The Agent.jl framework defines an abstract type called AbstractAgent. Each new custom agent type must possess an integer ID field. Agents situated in space also require a position field. The function responsible for advancing an agent's state is defined independently and should take a reference to the agent being advanced.

In the KrABMaga environment, the primary and most critical method is the step function, which determines the subsequent stage of the agent at the next time step. Mechanisms exist to invoke certain actions prior to and following the execution of the step function. Additionally, there is a method that determines if an agent should be removed from the queue after the current step. The AgentImpl structure serves as a wrapper for an Agent stored within the schedule, incorporating an ID field to differentiate agents.

In the Mesa framework, the Agent class possesses its own step function and includes a remove() method that facilitates the removal and deletion of the agent from the model. Identification numbers are associated with each agent and stored in the agent. These identification numbers are assigned by the model.

Within the MASON framework, an agent is represented as a class that implements the Steppable interface. The system incorporates a distinct step method and is utilized by several distinct StepMethod classes that implement the Steppable and Stoppable interfaces in various combinations for the user.

In Agent Harmony, an agent does not have to derive from any class. However, to be usable within the class *Schedule*, an agent should implement the *step* method, which is responsible for updating the agent's internal state and interacting with the environment, similarly to other frameworks.

2.4 Scheduler

In MASON, the Scheduler is a distinct class and serves as a property of the SimState model class. It has *step* method responsible for invoking update for every agent scheduled for a given timestamp. It stores *Steppables* in a priority queue, associated with a proper instance of class Key, which stores information about the time at which the agent update must be invoked, and the priority of an agent relative to the other agents scheduled at the same timestamp. To represent the behavior of agents updating themselves periodically, MASON provides the wrapper class *IterativeRepeat*, which implements *Steppable* and stores a reference to another *Steppable*. In its *step* method, it calls the identical method for the stored *Steppable*, and adds itself to the schedule again.

A similar approach to that used in MASON is applied in the krABMaga and Agent Harmony. In this case, these two frameworks are implemented almost identically. Both of them have agents stored in a priority queue. Which is sorted by the time and order of an agent's execution. Additionally, an object stored in the queue contains information about whether an event is repeating. If it is, then it is added to the queue again. The only difference between the two is that Agent Harmony allows events to be repeated less frequently than every step.

On the other hand, Mesa lacks an integrated internal scheduler. This functionality was replaced with the AgentSet functionality and the internal Model.steps property, which is used for time scheduling. The user decides when a given agent (or *AgentSet*) should invoke a function. The *do* method provided by an *AgentSet* makes invoking a certain function for all agents in the set easier.

In the Agents.jl framework, the scheduling mechanism is essentially a function that takes a model as an input parameter and produces an iterator of agent identifiers arranged by execution sequence. The framework provides some predefined, popular schedulers, such as random execution or execution in the order in which the agents are stored.

Schedules in Agents.jl and Mesa allow the sequential execution of updates for agents. It offers better time complexity of the execution – $O(n)$, where n is the number of agent, compared to the solutions based on the priority queue. Iteration through all agents stored on the priority queue is in the order of $O(n \cdot log(n))$. Additionally, sequential execution is more predictable than one from the priority queue. When agents are stored in some kind of array, agents added first to the model will execute their actions earlier. In the case of a priority queue based on a heap, the order of agents with the same priority is implementation-defined. However, schedules based on the priority queue offer a clearer interface for changing priorities between iteration steps. In Agent Harmony, MASON, and krABMaga, ensuring that the update of a given agent will execute before another comes down to ensuring that the user assigns it a higher priority when calling a scheduling function. When talking about functions (as in Mesa or Agents.jl) that only have access to the container with agents, and not to their priority order, the user has to remember to implement some sort of data structure to hold information about it.

2.5 Representation of Space

Space plays an essential role in simulation environments. Different kinds of spaces are used to represent real-world scenarios, e.g. hexagonal grid to represent clonal colony [14] or soil surface [11], a squared grid in Conway's Game of Life [2] or different kinds of networks [4,6]. In response to the high demand for diverse space representations, framework developers have provided numerous options for optimal modeling.

- **Grid (Discrete)**: The environment is discretized into square cells, with agents occupying and navigating among distinct points on a two-dimensional grid. Various frameworks offer two-dimensional implementations of this space, with MASON and Agents.jl extending to three dimensions. These spatial configurations may feature periodic boundaries. Typically, framework developers offer two grid types: one that permits only a single agent per cell at any given time, and another that allows multiple agents to occupy a cell concurrently.
- **Continuous Space**: Agents can navigate freely using continuous real-number coordinates, which facilitate smooth movement and interaction based on distance. Like the grid, it can have periodic boundaries and be two- or three-dimensional. However, higher dimensions are only provided by MASON and Agents.jl.
- **Network Space**: Agents situated at nodes of a graph can interact or travel through edges guided by the network's topology. In Mesa and Agents.jl, a network is a space in which agents are stored so they can move between nodes. In Agent Harmony, krABMaga, and MASON, the type of a network vertex is almost arbitrary and can represent an agent or a set of agents.
- **Hexagonal Grid**: This grid comprises hexagonal cells and offers more uniformity in neighboring distances and orientation. It is provided only by Mesa and MASON.

The Table 1 summarizes the space types offered by the framework.

Table 1. Comparison of Space Types in Different Frameworks

Space Type	KrABMaga	Mesa	MASON	Agent Harmony	Agents.jl
Grid 2D	Yes	Yes	Yes	Yes	Yes
Grid 3D	No	No	Yes	No	Yes
Continuous Space 2D	Yes	Yes	Yes	Yes	Yes
Continuous Space 3D	No	No	Yes	No	Yes
Network Space	Yes	Yes	Yes	Yes	Yes
Network Space (Vertex as Agent)	Yes	No	Yes	Yes	No
Hexagonal Grid	No	No	Yes	No	Yes

3 Implementation

The primary objective of this research was to evaluate which framework implementation demonstrates the highest level of efficiency, independent of the language and platform on which the framework was developed. This study facilitates a practical understanding of the most efficient solutions and provides an opportunity to develop a new framework that is optimally applicable. C++ was selected as the target language due to its superior performance capabilities [8], while taking into account compiler's optimization, compared to the alternative in terms of performance - Rust.

3.1 Implemented Functionalities

Each Agent-Based Modeling (ABM) framework accommodates a diverse array of methodologies tailored to each spatial configuration, facilitating the regulation of agents' behaviors during simulation processes. It is imperative that these procedures are executed with stability and are optimized to reduce memory usage and computational complexity. In intricate simulations, each method is typically invoked for every agent at each iteration of the step method, necessitating careful design and implementation to ensure scalability and performance [15].

To ensure the fair comparison between the frameworks, the intersection of the functionalities offered by each framework was selected. The only exception to this rule is the *get_neighborhood* and *get_neighbors* method offered by discrete space and *get_neighborhood* offered by network. This is because this method is one of the basic methods of agent modeling and is not implemented by krABMaga and MASON. However, MASON does not only implement *get_neighborhood* for the network.

In terms of space representation, the following types were selected: discrete space with a single agent on the field, discrete space with multiple agents on the field, continuous space, and a network in which each vertex represents a single agent.

For representing space with a single agent per cell, the implemented methods were selected based on the *Field* class from Agent Harmony, *GridSpaceSingle* from Agents.jl, *SingleGrid* from Mesa, and *DenseGrid2D* from MASON, as well as the similarly named class from krABMaga. Although the latter two classes are used to represent a space that allows multiple agents to be stored in a single location. However, they were chosen because a space containing a single agent at a given position can be represented using a space that allows multiple agents. These classes offer the following methods (not necessarily with identical names):

- *get_neighborhood* (not present in the krABMaga) - A method for retrieving the neighboring cells of the agent.
- *get_neigtbors* (not present in the krABMaga) - A method that returns all agents located in the neighboring cells of the agent.
- *move_agent* - A method for relocating the agent to another cell or position.
- *place_agent* - A method for positioning the agent at a designated location in space.
- *remove_agent* - A method for removing the agent from the simulation space.

An identical set of functionalities, also excluding *get_neighborhood* and *get_neighbors*, is offered by classes responsible for discrete space representations, allowing several agents to be stored in a single position, offered by the frameworks discussed.

Continuous space is represented by the *ContinuousSpace* class in Agent Harmony, an identically named class in Agents.jl and Mesa, *Continuous2D* in MASON, and *DenseGrid2D* in krABMaga. The common part of their functionality is very similar to that offered by the previously mentioned spaces and differs from them only in the absence of the *get_neighborhood* function. This time, the *get_neighbors* function is also defined for krABMaga.

For obvious reasons, classes representing a network offer a slightly different set of functionalities than those representing a two-dimensional space. The name *Network* for a class representing this type of space is used by MASON, krABMaga, and AgentHarmony. Mesa introduces a slight modification to it – *NetworkGrid*. Agent.jl, on the other hand, uses a completely different convention, calling this class *GraphSpace*. The functionalities offered by most frameworks for networks are as follows:

- *get_neighborhood* (not implemented by MASON and krABMaga) - A method that returns all agents adjacent to a given vertex.
- *addNode* - A method for adding a new node (agent) into the network.
- *removeNode* - A method for deleting a node (agent) from the network.
- *addEdge* - A method for establishing an edge between nodes.
- *removeEdge* - A method for removing an edge between nodes.

In addition, the schedule, as well as the base classes for the agent and model, have been implemented for each framework. In this case, however, we did not implement common functionalities, but rather tried to faithfully reproduce the philosophy of representation of these classes from the original. The motivation for this was the fact that these functionalities are responsible for the simulation process, so they have a common goal.

3.2 Implementation Details

Despite the fact that the frameworks discussed offer basic functionalities for the most popular types of spaces, the way these spaces are represented and the way these functionalities are implemented differ between them. Due to the fact that the implementation was done in C++, memory management and lifetimes, especially of agents, had to be taken into account, which may also affect their performance.

Model and Agent. MASON, as well as krABMaga, which is heavily inspired by MASON, do not require the class storing the simulation state to store agents. In this case, their lifetime is entirely dependent on the scheduler, which is why, in our implementation, MASON operates on shared pointers that are passed to the scheduler, which stores them in its internal structures. KrABMaga, on the other hand, operates on unique pointers that are stored in the *AgentImpl* class used by the scheduler. For this reason, the simulation space contains copies of agents, not pointers to them. The Agent.jl implementation uses a map to store agents. This eliminates the potential for copying elements when the vector size changes.

The map maps the agent ID to the actual object. This object should have an ID as well as a pos variable, which is the agent's position in space. Additionally, the model is templated by the type of function that performs a step for the agent, as well as the schedule. In the case of Agent Harmony, the *Model* class stores agents as tuples of *std::deque* class objects, where each *std::deque* stores agents of different types. Thanks to this approach and the inability to remove an agent from the model, Agent Harmony guaranties the constancy of agent pointers.

Single Agent Grid. The implementation of single agent space is more or less consistent across all frameworks. It is a two-dimensional array of specific sizes that stores a specific type of data. In Agent Harmony, it is an optional variant of indicators for all agents appearing in the simulation. In the case of Mesa, the cell corresponds to a shared pointer to the *Agent* class, which is the base for all agents in the simulation. The only exception to storing a pointer to an agent in this space representation occurs in Agent.jl. The space in this framework is still represented by a two-dimensional array, but it stores the ID of the corresponding agent at its position. If there is no agent at a given position, it takes the value 0.

Multi Agent Grid. The implementation of multi-agent space is very similar to the implementation of single-agent space. In all frameworks, it is also a two-dimensional array, and each position contains a vector. The only difference is in the type stored in this vector. Similar to a grid with a single agent per position, Mesa and AgH store, respectively: a shared pointer to the *Agent* class and a variant (this time not optimal) of pointers to all possible agent types that can be stored in this space. MASON originally stores instances of the *Object* class in this "vector," so in our implementation, the *DenseGrid2D* class is templated with the type of object stored in this space and stores a pointer to that type in the vector. It is also worth mentioning that MASON uses its own type of vector, called *Bag*, in which removing a given element involves swapping it with the last object in the vector and then removing the last element. Agents.jl once again stores the agents associated with a given position at the position id. It uses a similar model to MASON for removing an agent from a given position. KrABMaga is the only one that does not store any references to the object, but rather the actual object (or more precisely, a copy of it). In the case of larger agents, this may involve additional time needed to copy the object in order to place it in the appropriate location.

Continuous Space. The main approach encountered during the implementation of continuous space is its discretization. This approach is used by Agents.jl, krABMaga, and Agent Harmony. The framework written in Julia stores the IDs of the relevant agents, krABMaga stores a copy of the object, and Agent Harmony stores a variant of indicators. MASON uses a very similar implementation, but it does not discretize continuous space into a two-dimensional array; instead, it discretizes it into a map where the key is the discretized position of the agent. A completely different implementation is used in the case of Mesa. In this case, continuous space is represented by a vector of shared pointers. However, this implementation means that when retrieving neighbors, the distance from each agent present in the space must be calculated. It is worth mentioning that MASON is the only framework that does not require the agent to store information about its position; this is done by a map located in the *Continuous2D* class, which maps the agent's pointer to its position in space.

Schedule. For Mesa and Agents.jl, the schedule was not implemented because, for both of these frameworks, the function is responsible for the correct order of actions. In the case of Mesa, this is the *step* method of the *Model* class. In the case of Agents.jl, it is a function that accepts a model and returns a vector with agent IDs. The implementation of schedulers for krABMaga, MASON, and Agent Harmony is nearly identical. The class *std::priority_queue* is used to represent the queue. The objects stored in the queue contain, respectively: a reference to the agent that is to be executed; in the case of Agent Harmony, this is a variant of pointers; for krABMaga, a unique pointer to the *Agent* class; and for MASON: a shared pointer to *Steppable*; a key by which the elements are sorted; this key consists of the time at which the agent is to be executed and the priority relative to other agents queued for the same moment in time.

4 Benchmarks

The algorithms were evaluated for each spatial configuration. A comprehensive evaluation was conducted where each performance test was done for different grid sizes and the schedule test was done for different numbers of agents. A series of tests were conducted to assess the performance of the algorithms under various conditions. The time needed to complete each test was measured and recorded.

4.1 Methodology

Agent-based modeling (ABM) frameworks typically encompass various components, including schedulers, step functions, and agent containers. However, space operations—such as querying neighbors, moving agents, or spatial indexing—are among the few universally present and consistently executed components across all platforms examined, including KrABMaga, Mesa, MASON, Agents.jl, and Agent Harmony (C++). Conversely, other elements, such as schedulers or the step() function, exhibit significant variation in their design, usage patterns, and even presence across these frameworks. This heterogeneity renders them unsuitable for systematic, apples-to-apples benchmarking or for establishing a consistent cross-framework mapping in C++. In contrast, space operations are foundational and consistently accessible, rendering them the most appropriate candidates for comparative analysis. These operations constitute the computational foundation of most agent interactions, irrespective of the higher-level modeling logic. Concentrating on space operations enabled performing rigorous and equitable experiments and served as a robust foundation for developing a generalized mapping layer within the C++ Agent Harmony framework.

To ensure identical test conditions for each framework, the pseudo-random number generator (*std::mt19937*) was initialized with the same seed.

In the case of testing the Single Agent Grid, agents were randomly placed on every other cell of the space. Then, each agent was randomly moved to an empty location. The next task was to retrieve neighboring cells within a radius of 2 and 3 for each agent. This task was skipped for krABMaga due to the lack of a *get_neighborhood* function. Next, for each agent, all neighbors within the same radius were retrieved. This step was also skipped for krABMaga. Finally, each agent was removed from the space. The times for each of these actions were measured and recorded in a csv file.

The same tasks as in the case of the Single Agent Grid were performed for the Multi Agent Grid. Again, for the same reasons, neighbor retrieval was skipped for krABMaga. The only difference was the placement of a number of agents five times greater than the number of cells in the grid. The Single Agent Grid and Multi Agent Grid tests were performed for grids of sizes: 64, 128, 256, and 512.

The task set for continuous space consisted of adding one agent to each square unit of the map. In those frameworks where space discretization was required, this value was set to 1. The set of tasks performed after adding agents to the space was analogous to that in discrete spaces. First, agents were moved to a random position within the grid, then neighboring agents within a radius of 2 and 3 were retrieved for each agent, and finally, all agents were removed from the space one by one. Tests on continuous space were run for squared grids of sizes: 64, 128, 256, and 320.

The performance test for space represented using a network involved adding a specified number of nodes (agents) to the network. These nodes were then randomly connected to each other. The next action was to retrieve all neighbors within a radius of 1, 2, and 3, respectively. The final two tasks involved removing all connections between the nodes, followed by removing the nodes from the network. The tests were performed for networks with the following number of nodes: 64, 128, 256, and 512. Each node was connected to $\lfloor \frac{n}{10} \rfloor$ other nodes, where n is the number of nodes in the whole graph.

The scheduler test involved creating a model capable of holding 4 different types of agents and running a simulation on it. The same number of agents of each type was added to the model each time. The step for each agent consisted of incrementing a counter stored in the model by a random value in the range from 1 to 10.

- Type 1 agents executed actions every step with priority 1.
- Type 2 agents also acted every step, with priority 2.
- Type 3 agents acted every 3 steps, with priority 1.
- Type 4 agents acted every 5 steps, also with priority 1.

The time for the initial scheduling of agents was measured separately from the simulation runtime. The test was run for 1024 iterations, with 32768, 65536, and

131072 agents, respectively. These values were selected to see how the time was scaling with the number of agents and also to avoid ending the simulation almost instantly (Table 2).

Table 2. Test parameters. Superscripts denote: [1]number of cells per row/column, [2]width/height of space, [3]number of nodes.

Tested component	Space sizes	No. of agents	Tested radii	No. of iterations
SingleGrid	4, 128, 256, 512[1]	2048, 8192, 32768, 131072	2, 3	–
MultiGrid	4, 128, 256, 512[1]	2048, 8192, 32768, 131072	2, 3	–
Continuous space	4, 128, 256, 320[2]	4096, 16384, 65536, 102400	2, 3	–
Network	4, 128, 256, 512[3]	4, 128, 256, 512	1, 2, 3	–
Scheduler	–	2768, 65536, 131072	–	1024

4.2 Speed Performance

The figures show the average execution time determined from multiple iterations of test trials.

Operations on Singlegrid. As demonstrated in Fig. 1, Agent Harmony displays the most expedited performance in incorporating agents into the grid, particularly excelling in scenarios with smaller grid dimensions. Nonetheless, in expansive grid environments, Mesa appears to parallel the performance of Agent Harmony. In the context of agent removal and movement across grids, Agents.jl emerges as the most efficient solution, outperforming other frameworks considerably, especially in larger spaces, surpassing KrABMaga and Agent Harmony by approximately 421% and 360%, respectively. Notably, as evidenced in Fig. 2, KrABMaga's algorithm exhibits superior efficacy compared to Agents.jl when processing mid-sized grids. As depicted in Figs. 3a and 3b, Agent Harmony consistently delivers the highest performance across all examined scenarios. MASON follows as the second most effective, although it notably experiences a decrease in efficiency in smaller grid contexts.

Operations on Multigrid. Analogously to the SingleGrid methods, in the cases of *get_neighborhood* and *get_neighbors*, Agent Harmony achieved the most efficient and rapid execution time for a single operation, as illustrated in Figs. 6a and 6b. Remarkably, MASON ranks second, with Mesa demonstrating inferior results compared to all other frameworks. Regarding the processes of adding, removing, and moving agents in a multigrid space, all frameworks exhibited comparable performance outcomes. Notably, however, the MASON implementation appears to perform slightly better than other implementations of these methods, as depicted in Figs. 4b, 7a, and 5.

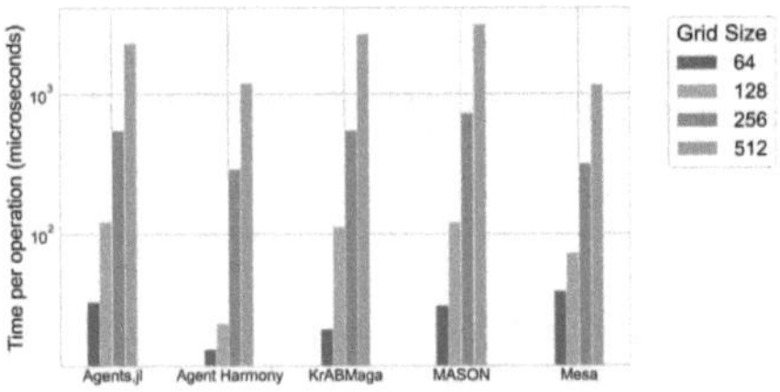

(a) Agents adding operation time in SingleGrid space for all frameworks

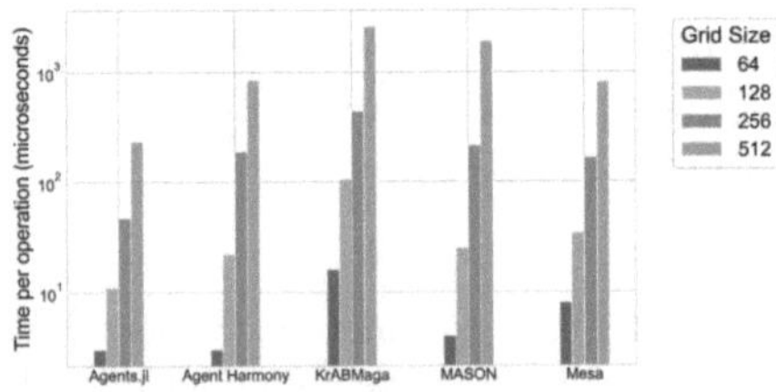

(b) Agents moving operation time in SingleGrid space for all frameworks

Fig. 1. Times for adding and moving agents in SingleGrid space

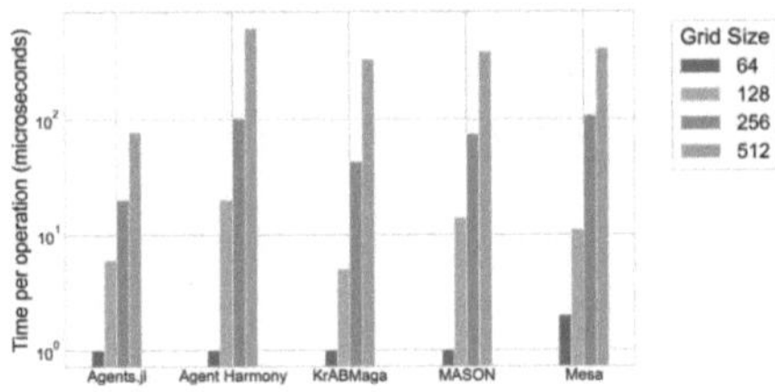

Fig. 2. Agents removing operation time in SingleGrid space for all frameworks.

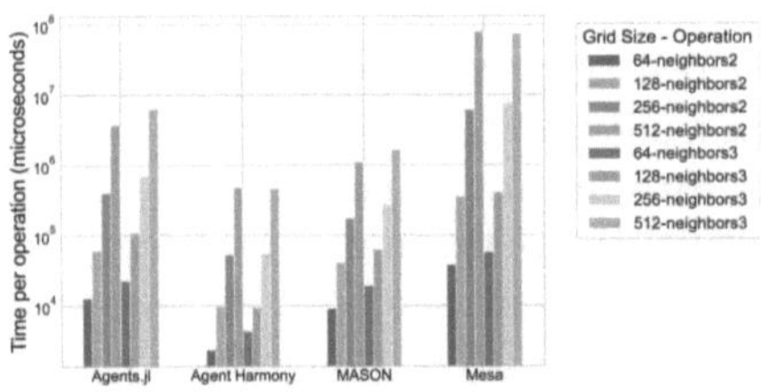

(a) Agents get_neighbors operation time in SingleGrid space for all frameworks.

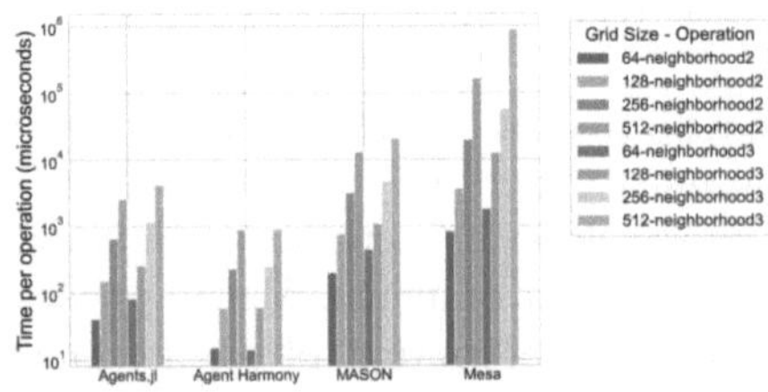

(b) Agents get_neighborhood operation time in SingleGrid space for all frameworks.

Fig. 3. Times for get_neighbors and get_neighborhood in SingleGrid space

Operations on Continous Space. As depicted in Fig. 7a, the integration of agents into the ContinuousSpace grid presents a complex challenge. Performance is of utmost importance in this context, given the broad application of continuous space in numerous large-scale, real-world agent-based simulations [13]. Mesa, KrABMaga, and Agent Harmony demonstrate comparable performance capacities in this domain. Conversely, operations involving movement and removal exhibit significant variation. Mesa markedly surpasses all other frameworks in movement operations, as illustrated in Fig. 8a. This is particularly critical since the method referenced in *move* is frequently employed during simulations. Conversely, Mesa exhibits significantly inferior performance in agent removal, as shown in Fig. 7b. Agents.jl and Agent Harmony exhibit superior efficiency in

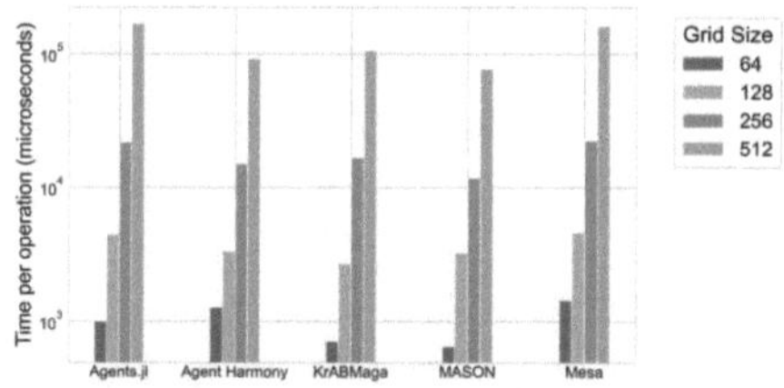

(a) Agents adding operation time in MultiGrid space for all frameworks

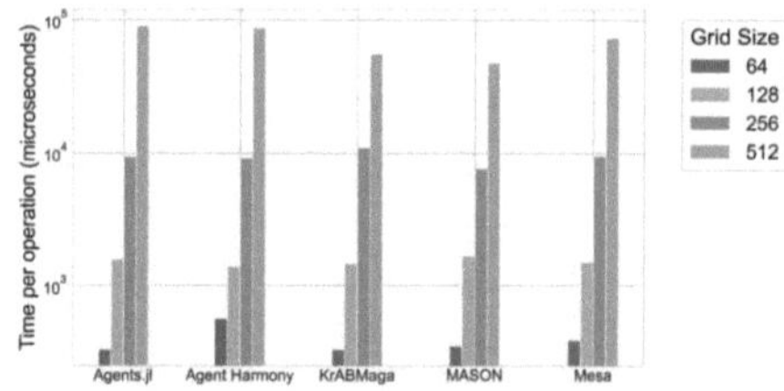

(b) Agents removing operation time in MultiGrid space for all frameworks

Fig. 4. Times for adding and removing agents in MultiGrid space

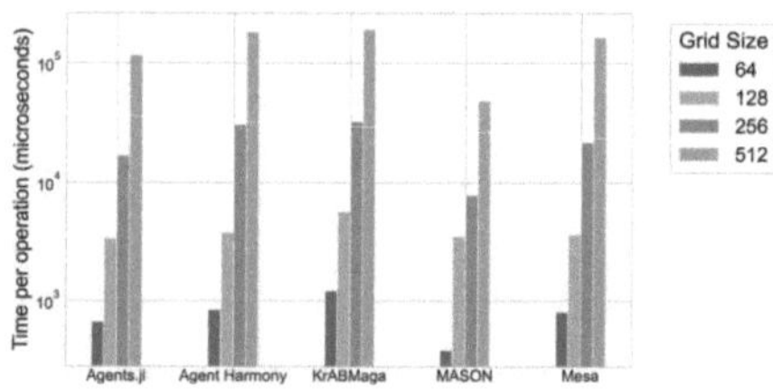

Fig. 5. Agents moving operation time in MultiGrid space for all frameworks

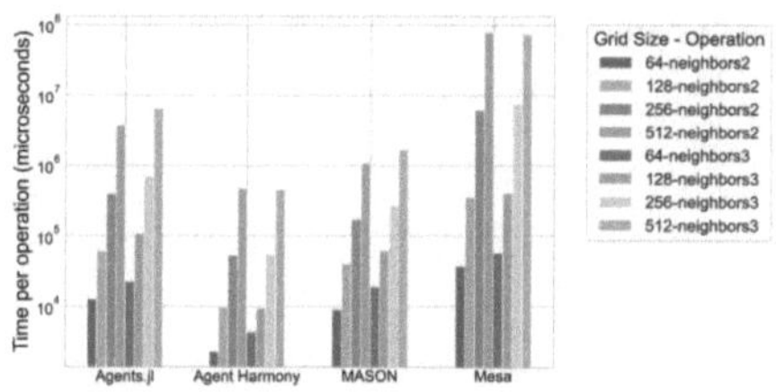

(a) Agents get_neighbors operation time in MultiGrid space for all frameworks

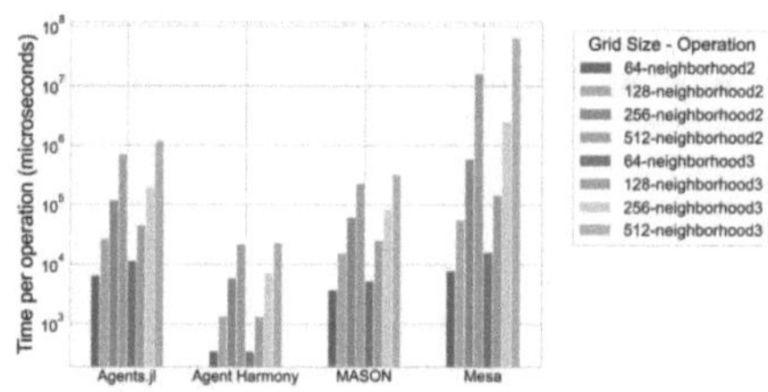

(b) Agents get_neighborhood operation time in SingleGrid space for all frameworks

Fig. 6. Times for get_neighbors and get_neighborhood in MultiGrid space

agent removal, with Agents.jl excelling in larger grids, and Agent Harmony showing better performance in smaller ones. Regarding neighborhood and neighbor acquisition, KrABMaga and Agent Harmony are the leading performers, with Agent Harmony showing slightly superior results, as evidenced in Fig. 8b.

Operations on NetworkGrid. In the context of NetworkGrid, among the tested algorithms and implementations, Agents.jl demonstrated superior efficacy in identifying neighbors, as illustrated in Fig. 11. Regarding node connectivity, all frameworks exhibited similar performance, with KrABMaga and MASON leading, as presented in Fig. 10a. For adversarial operations, KrABMaga achieved the most favorable outcomes. Notably, Agent Harmony, Mesa, and Agents.jl

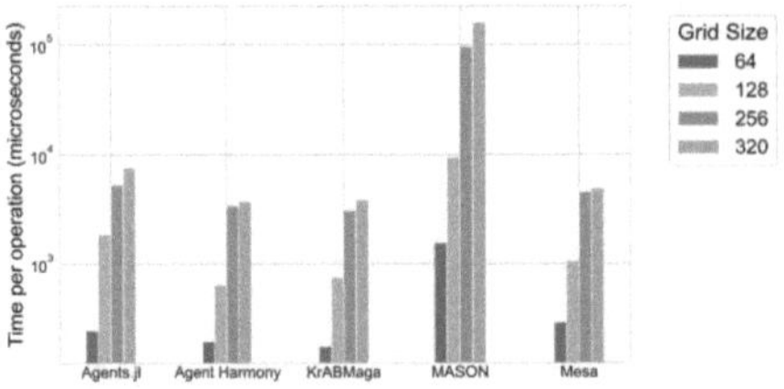

(a) Agents adding operation time in Continuous space for all frameworks

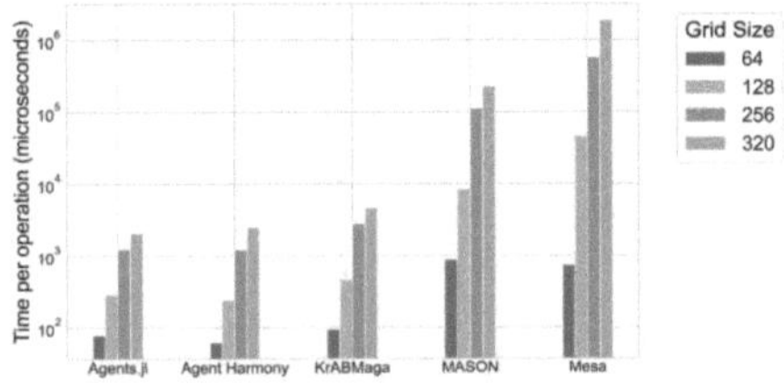

(b) Agents removing operation time in Continuous space for all frameworks

Fig. 7. Times for adding and removing agents in Continuous space

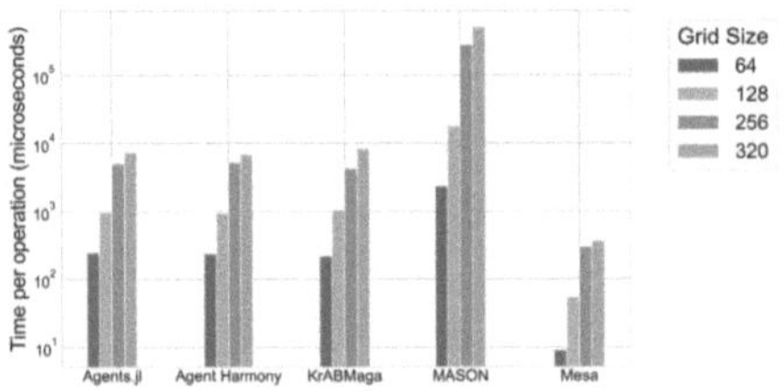

(a) Agents moving operation time in Continuous space for all frameworks

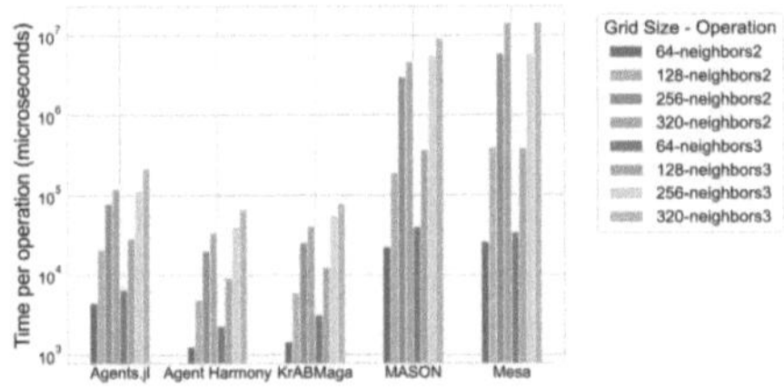

(b) Agents get_neighbors operation time in Continuous space for all frameworks

Fig. 8. Times for moving agents and getting its neighbors in Continuous space

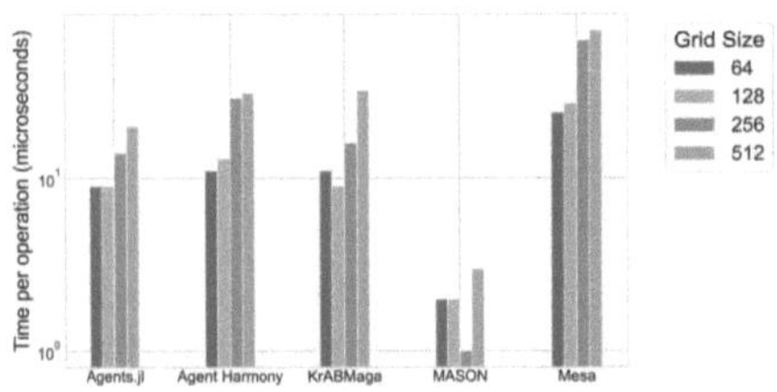

(a) Agents adding operation time in Network for all frameworks

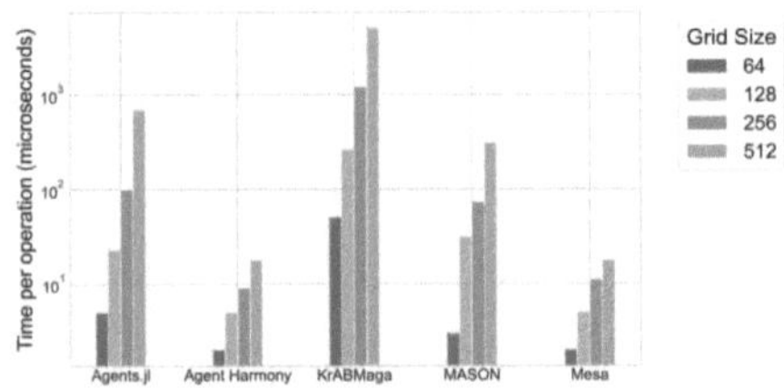

(b) Agents removing operation time in Network for all frameworks

Fig. 9. Times for adding and removing agents in Network.

exhibited comparable performance levels, while MASON underperformed significantly, as shown in Fig. 10b. With respect to node operations, the addition of new nodes to the network was most efficiently executed by MASON, with all other frameworks falling substantially short of expectations, as detailed in Fig. 9a. In contrast, node removal was performed most effectively by Mesa, followed closely by Agent Harmony, with only a slight disparity between them. Other frameworks exhibited markedly lower performance, as depicted in Fig. 9b.

Scheduling Operations. Various frameworks adopt distinct methodologies for scheduler implementation, employing diverse data structures for event storage

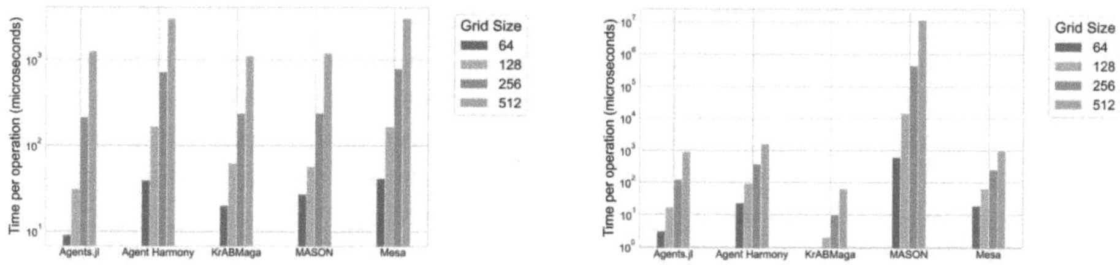

(a) Connecting nodes in network operation time for all frameworks

(b) Unconnecting nodes in network operation time for all frameworks

Fig. 10. Times for connecting and unconnecting nodes in Network

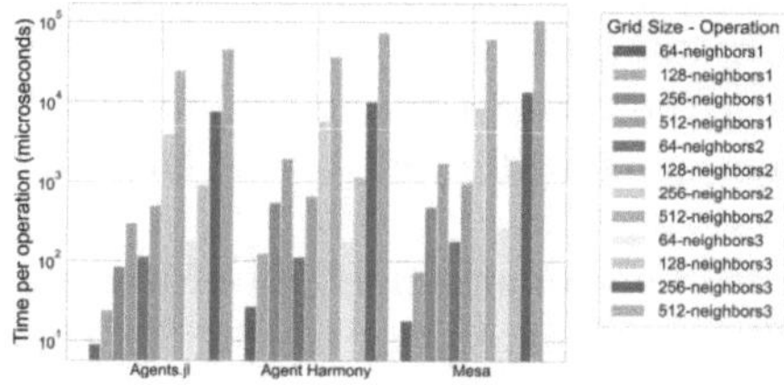

Fig. 11. Agents get_neighbors operation time in Network for all frameworks

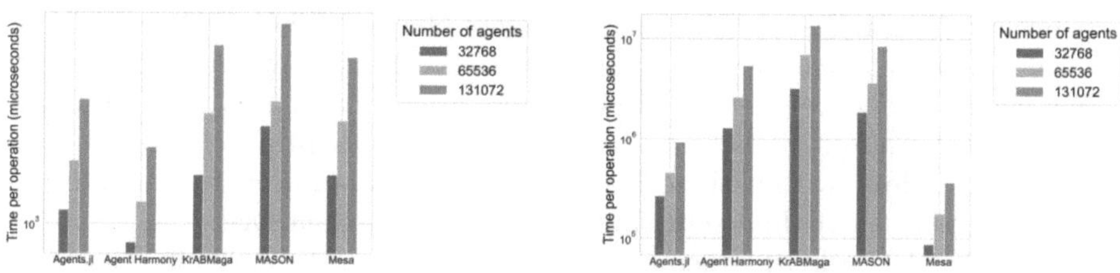

(a) Adding events to Scheduler operation time for all frameworks

(b) Running list of events by Scheduler total time for all frameworks

Fig. 12. Times for adding agents and running Scheduler

and varying queue mechanisms for event scheduling. In the context of this research, an event is conceptualized as an Agent. As depicted in Fig. 12a, Agent Harmony executes the process of adding new agents to the queue with optimal efficiency, significantly exceeding other frameworks. Regarding the execution of agents' actions, the Mesa implementation yielded the most favorable results, with Agents.jl trailing as a distant second, as illustrated in Fig. 12b.

5 Conclusions

- In the Single Agent Grid context, Agent Harmony excels in neighborhood and neighbor retrieval, thanks to its effective spatial indexing algorithms. Conversely, Agents.jl demonstrates the quickest agent movement and removal in larger grids due to its efficient memory management. Although KrABMaga does not support neighborhood functionality, it still handles movement and removal efficiently in mid-sized grids. Mesa consistently performs well across various operations and competes closely with Agent Harmony in agent addition, notably in large grids. While MASON trails a bit in smaller grids, it offers reliable performance in neighborhood queries.

- In Multi Agent Grid tests, Agent Harmony performs better for neighborhood and neighbor lookups, with a reaffirmation of scalability and speed of its spatial querying methods. MASON slightly edges out competitors in the performance of basic space operations—adding, removing, and moving agents—highlighting the robustness of its internal data structures. Although Mesa trails behind in neighborhood queries, it maintains acceptable performance overall. Notably, all frameworks show minimal disparity in basic operations in this scenario, with the competitive edge shifting based on specific task demands. KrABMaga is again limited by its lack of neighbor-related functionality.
- The Continuous Space benchmark highlights Mesa's superiority in terms of agent movement efficiency, thanks to its efficient continuous space navigation algorithms. However, Mesa shows considerable latency in agent removal, which could impact simulations involving frequent deletions. Agents.jl is particularly adept at removal tasks on larger grids, likely due to optimized vetor, which stores only agents' ids, whereas Agent Harmony delivers balanced performance in both movement and removal activities, notably in smaller grids. For neighbor and neighborhood detection, Agent Harmony and KrABMaga are at the forefront, indicating the presence of well-optimized spatial proximity algorithms within these frameworks. This area accentuates the vital importance of algorithmic efficiency in real-time simulations.
- Within the NetworkGrid domain, Agents.jl showcases the most efficient neighbor discovery, benefiting from its high-performance graph traversal techniques. KrABMaga and MASON outperform in connection operations, likely owing to their optimized data structures for handling graph edges. Conversely, Agent Harmony and Mesa excel in edge removal, though MASON experiences a decline in performance for this task. Node insertion is most efficiently managed by MASON, possibly due to its lightweight graph structure. Mesa leads in node deletion performance, with Agent Harmony close behind, suggesting effective memory and index management. Each framework contributes distinct strengths to network-based simulations.
- The performance evaluation reveals significant differences in scheduling techniques across frameworks. Utilizing a priority queue within a specific scheduler class, Agent Harmony excels in efficiently adding agents to the schedule. Conversely, Mesa implements a step method for agent execution within the Model class, resulting in the most effective performance during agent action runs. Although Agents.jl comes next in execution efficiency, it considerably trails behind. These findings highlight the influence of data structures and architectural decisions on the comprehensive performance of agent-based simulations.

6 Future Work

The study's comprehensive examination and implementation of algorithms, techniques, and spatial structures from diverse Agent-Based Modeling Frameworks in C++ establishes a robust foundation for future research. Firstly, the benchmarking outcomes can be a basis for more detailed performance analysis and cross-language evaluations, allowing the community to gain a deeper understanding of the trade-offs among different agent-based modeling systems. Secondly, the modular architecture of the implemented components facilitates iterative enhancements and optimizations for particular methods like neighbor retrieval or agent scheduling, directly boosting the performance of current frameworks. Furthermore, an understanding of the strengths and weaknesses of each system paves the way for creating a new hybrid framework that efficiently integrates top-performing algorithms, data structures, and design concepts from Agent Harmony, Agents.jl, Mesa, MASON, and KrABMaga. This hybrid model might offer both superior performance and wide-ranging capability, establishing a new benchmark for scalable and adaptable agent-based simulations. This work also inspires further investigation into optimizations specific to certain domains and the fusion of GPU acceleration, parallel processing, or adaptive data structures to expand the capabilities of agent-based modeling.

Acknowledgments. The research was financed by a scientific subsidy from the AGH University of Krakow.

Disclosure of Interests. The authors declare that they have no known competing financial interests or personal relationships that could have appeared to influence the work reported in this paper.

References

1. Antelmi, A., Cordasco, G., D'Auria, M., De Vinco, D., Negro, A., Spagnuolo, C.: On evaluating rust as a programming language for the future of massive agent-based simulations. In: Tan, G., Lehmann, A., Teo, Y.M., Cai, W. (eds.) AsiaSim 2019. CCIS, vol. 1094, pp. 15–28. Springer, Singapore (2019). https://doi.org/10.1007/978-981-15-1078-6_2
2. Conway, J., et al.: The game of life. Sci. Am. **223**(4), 4 (1970)
3. Datseris, G., Vahdati, A.R., DuBois, T.C.: Agents.jl: a performant and feature-full agent-based modeling software of minimal code complexity. Simulation **0**(0), 003754972110688 (2022). https://doi.org/10.1177/00375497211068820
4. de Durana, J.M.G., Barambones, O., Kremers, E., Varga, L.: Agent based modeling of energy networks. Energy Convers. Manage. **82**, 308–319 (2014)
5. El-Sayed, A.M., Scarborough, P., Seemann, L., Galea, S.: Social network analysis and agent-based modeling in social epidemiology. Epidemiol. Perspect. Innov. **9**(1), 1–9 (2012)

6. Frias-Martinez, E., Williamson, G., Frias-Martinez, V.: An agent-based model of epidemic spread using human mobility and social network information. In: 2011 IEEE Third International Conference on Privacy, Security, Risk and Trust and 2011 IEEE Third International Conference on Social Computing, pp. 57–64. IEEE (2011)
7. Gacek, P., et al.: A comparison of selected agent-based modelling frameworks. In: International Conference on Parallel Processing and Applied Mathematics, pp. 276–288. Springer (2024)
8. Ivanov, N.: Is Rust C++-fast? Benchmarking system languages on everyday routines. arXiv preprint arXiv:2209.09127 (2022)
9. Kazil, J., Masad, D., Crooks, A.: Utilizing Python for agent-based modeling: the Mesa framework. In: Thomson, R., Bisgin, H., Dancy, C., Hyder, A., Hussain, M. (eds.) Social, Cultural, and Behavioral Modeling, pp. 308–317. Springer International Publishing, Cham (2020)
10. Khodabandelu, A., Park, J.: Agent-based modeling and simulation in construction. Autom. Constr. **131**, 103882 (2021)
11. Kim, M., Or, D.: Individual-based model of microbial life on hydrated rough soil surfaces. PLoS ONE **11**(1), e0147394 (2016)
12. Luke, S., et al.: The MASON simulation toolkit: past, present, and future. In: Multi-Agent-Based Simulation XIX: 19th International Workshop, MABS 2018, Stockholm, Sweden, 14 July 2018, Revised Selected Papers, vol. 19, pp. 75–86. Springer (2019)
13. Macal, C.M., North, M.J.: A tutorial on agent-based modeling and simulation. In: Proceedings of the 37th Conference on Winter Simulation, pp. 2–15 (2005)
14. Mony, C., Garbey, M., Smaoui, M., Benot, M.L.: Large scale parameter study of an individual-based model of clonal plant with volunteer computing. Ecol. Model. **222**(4), 935–946 (2011)
15. RailsBack, S.F., Grimm, V.: Agent-Based and Individual-Based Modelling: A Pracitcal Introduction. Princeton Univeristy Press (2019)
16. Wrona, Z., et al.: Comparison of multi-agent platform usability for industrial-grade applications. Appl. Sci. **14**(22) (2024). https://doi.org/10.3390/app142210124. https://www.mdpi.com/2076-3417/14/22/10124
17. Ws, J., Lubaś, R.: Towards realistic and effective agent-based models of crowd dynamics. Neurocomputing **146**, 199–209 (2014)
18. Zięciak, A.: Agent Harmony (2025). https://github.com/Tosiekdev/Agent-Harmony

Intelligent Agents in Biomedical Applications

An Empirical Comparison of Unimodal and Multimodal Breast Cancer Detection

Betul Kantaroglu and Jing Hua Ye(✉)

Computer Science, Munster Technological University, Bishopstown, Cork City, Republic of Ireland
betul.kantaroglu@mymtu.ie, jinghua.ye@mtu.ie

Abstract. Among many types of cancer, breast cancer remains one of the leading causes of cancer-related deaths worldwide. Early diagnosis is crucial for survival rates. This study aims to empirically investigate the performance and limitations of various machine learning and deep learning techniques applied to both unimodal and multimodal breast cancer datasets. Unimodal models (such as clinical metadata or mammograms) serve as baselines, allowing for comparisons with the multimodal models explored in this study. Multimodal models were trained using various techniques, with transfer learning applied to both mammograms and clinical data as the baseline. Unimodal classifiers (MLP Neural Network, Random Forest, Logistic Regression, XG-Boost) across three different datasets (WDBS, WBC, and BCSC) served as fundamental baselines, providing a benchmark for evaluating the effectiveness of more complex multimodal approaches. The fundamental baseline for the multimodal model was the transfer learning model, which achieved an accuracy of 0.8148 and an F1-score of 0.81. Our study revealed that unimodal models (baselines) trained solely on clinical metadata outperformed the multimodal models. The baseline model for the multimodal experiments was only successful in outperforming the Logistic Regression performed on the BCSC dataset (achieved an accuracy of 0.69).

Keywords: Multimodal Breast Cancer · Unimodal Breast Cancer · Transfer Learning

1 Introduction

Breast cancer is a disease in which abnormal breast cells grow out of control and form tumors. These tumors can spread throughout the body and can result in fatality if left untreated. Breast cancer remains one of the most common cancers among women worldwide and contributes significantly to cancer-related deaths. According to the World Health Organization (WHO), breast cancer took the lives of 670,000 patients in 2022 alone [1]. In Ireland, 3,600 women and approximately 30 men are diagnosed with breast cancer each year [2]. Early detection and accurate classification are crucial for enhancing patient outcomes,

© The Author(s), under exclusive license to Springer Nature Switzerland AG 2026
K. Mason et al. (Eds.): IASE 2025, CCIS 2976, pp. 69–82, 2026.
https://doi.org/10.1007/978-3-032-28263-7_5

enabling timely treatment, and reducing mortality rates. Breast cancer survival rates are increasing, and breast cancer accounts for 22% of cancer survivors [3].

With the rise of digital healthcare and access to large datasets, machine learning (ML) and deep learning (DL) have shown promise in assisting breast cancer diagnosis. Traditional diagnosis relies on single-modality information, such as mammographic images or clinical features, limiting the scope and accuracy of prediction. Although previous research has yielded strong results using individual data types, such as mammograms or clinical variables, combining different sources of information can provide a more comprehensive picture of a patient's condition.

The primary motivation behind this study is the potential to combine multiple data modalities, including clinical records, imaging, and textual information, to improve diagnostic accuracy. We will create a classification system capable of working with various types of data and assess its performance under different conditions. This system facilitates breast cancer classification using either an unimodal ML approach or a multimodal DL approach. It integrates clinical data, imaging features, and simulated textual patient history. A variety of models were developed, and their performance was evaluated, including logistic regression (LR), support vector machine (SVM), EfficientNet for visual input, and transformer-based models for text. The transfer learning (TL) technique was utilized to improve generalization [8].

In the upcoming section, we will review some existing approaches and techniques carried out in this field. The details of building this framework are discussed in Sect. 4. The performance of this framework is analyzed in Sect. 5. We will conclude our study with our research findings and some possible future research directions.

2 Literature Review

In this section, we will provide some general background on breast cancer and will review some cutting-edge approaches and techniques applied to detect this disease.

2.1 Breast Cancer Overview

Breast cancer, a heterogeneous disease, is one of the most common forms of cancer affecting women worldwide. It occurs when abnormal cells in the breast tissue begin developing and growing at an uncontrollable rate, forming tumors. In many cases of late diagnosis, the tumor may invade surrounding tissues or organs. Most commonly, it develops in the breast tissue, often in the ducts (which are channels responsible for carrying milk from lobules to the nipple) or in the lobules themselves (glands that produce milk). While breast cancer primarily affects women, it is important to note that breast cancer can also be diagnosed in men, in rare cases [3] (Fig. 1).

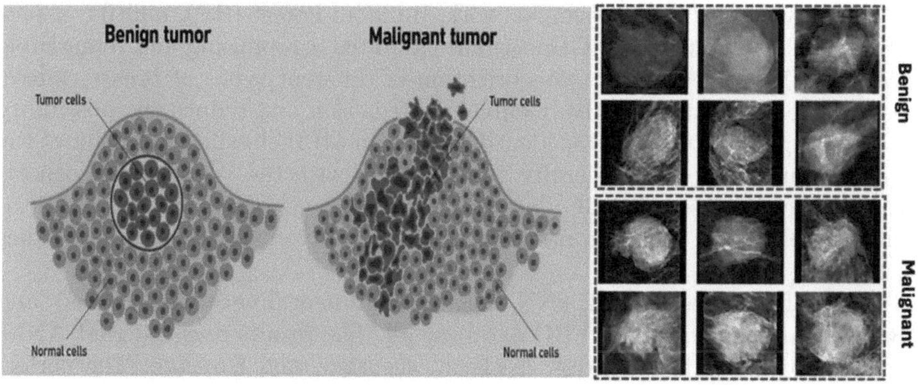

Fig. 1. The visualization of tumour cells and tumour masses for the benign and malignant cases of breast cancer [7].

Breast cancer is present in various subtypes, each with different biological characteristics, responses to treatment, and outcomes. Early detection plays a vital role in increasing the chances of successful treatment and reducing the risk of spreading to other parts of the body [13]. That is why a major emphasis on screening programs, regular checkups, and advances in diagnostic technology is encouraged. The most common symptoms, like a sudden lump or thickening in or near the breast, etc., do not always indicate cancer; however, any changes should be immediately evaluated by a qualified healthcare professional. Breast Self-Examination (BSE) does not replace the evaluation of a medical professional; however, it can provide individuals with a sense of familiarity with the normal look and feel of their breasts. Frequently performing self-examinations can help catch any changes that may have occurred (lumps, skin dimpling, skin scaling, discharge from the nipple, changes in size or shape). BSE is recommended to be performed once a month, ideally at the end of the menstrual cycle (for women), when breasts are least likely to be swollen or tender.

Before the introduction of AI and advanced ML methods, breast cancer diagnosis relied almost entirely on manual processes and clinical expertise, for instance, clinical breast examination, mammography, ultrasound, biopsy, histopathological examination, and risk assessment models [9]. Undoubtedly, these traditional diagnostic methods have saved countless lives and are still used widely; however, they have several limitations – human errors, dense breast tissue, or subtle abnormalities can confuse and result in misdiagnosis, be time-consuming, and cause data overload. These challenges highlight the importance of integrating new AI technologies into breast cancer diagnosis.

2.2 Unimodal Breast Cancer Detection

ML techniques have been intensively studied and widely utilized to detect breast cancer through imaging or genomics. The advent of various ML techniques, like the Naïve Bayes (NB), Bayesian Network, Random Forest (RF), multilayer perception (MLP), SVM, and XGBoost, has exhibited a promising and remarkable success in the detection and categorization of different types of breast cancer. Ensemble learning can effectively enhance prediction and reduce the possibility of overfitting [4]. Literature has identified that an MLP classifier can achieve the highest accuracy 78.6% to identify breast cancer with the Wisconsin Prognosis Breast Cancer dataset (WPBC) [5]. SVM, k-nearest neighbors (kNN), and decision trees (DTs) are the main methods predominantly employed in prior studies [4].

The performance of SVM, kNN, RF, artificial neural networks (ANNs), and LR was examined with the WBC database [6]. The results revealed that ANNs and SVM achieved the greatest precision, accuracy, and F1-score. The performance of RF, kNN, and NB was compared with the WDBC dataset [10]. An ML-based image processing approach [11] was presented to identify breast cancer with the MIAS dataset. In this model, the quality of an image was improved with a geometric mean filter, an AlexNet was used to extract features, and features were picked with the relief algorithm. The cancer was categorized with one of the following models: SVM, kNN, RF, and NB. The DT and XGBoost classifiers achieved the best accuracy of 97% and the highest AUC of 0.9999 compared to other models [12].

2.3 Multimodal Breast Cancer Detection

Numerous studies have been conducted on predicting breast cancer with images and genetic information. Prediction accuracy can be improved with multimodal approaches that combine images, clinical information, and molecular data [14]. A multimodal DL model [15] was built to classify the subtypes of breast cancer. This model integrates breast mammography images with clinical metadata features, like age and the class of a tumor. This model extracts all features with a CNN. The experimental results demonstrate that the AUC of this model (88.87%) outperforms the AUC of an unimodal model based solely on mammography images (61.3%). A novel framework, MM-SurvNet [16], was designed to improve risk prognostication of breast cancer. In this framework, all features are extracted with the MaxViT model. Genetic data were fused with images using a dual cross-attention mechanism; the clinical data are incorporated in the final layer. A deep multimodal learning framework [17] was proposed to improve the prediction of breast cancer; this framework is a fully connected neural network. In this framework, image features were extracted from mammograms with the ResNet50 model, and those extracted features were fused with clinical data using CNN; TL was employed to reduce training time by picking up pre-learned image representations. This framework obtained 90.8% accuracy.

2.4 Remarks

The existing studies mainly focused on developing a new model to improve the performance of breast cancer detection in terms of accuracy, AUC, F1 score, recall, and precision. K-fold cross-validation is the main validation method used to assess the performance of a newly developed model in these existing studies. The performance of the classifiers was empirically compared with either the unimodal dataset or the multimodal dataset. We have not found any studies empirically comparing the performance of classifiers with the unimodal dataset and the performance of classifiers with the multimodal dataset. Furthermore, the batch validation method is applied instead of the conventional k-fold cross-validation method. In this study, we are inclined to answer the following research question: Can combining image and textual clinical information improve the performance of breast cancer classification compared to an unimodal approach?

3 Preliminary Unimodal and Multimodal Breast Cancer Detection Framework

The system architecture is designed to integrate and process several heterogeneous data sources to provide a comparative platform for breast cancer classification. As shown in Fig. 2, the system is multimodal (unimodal models are also facilitated for serving as the baseline for performance comparisons) with different pipelines for image inputs (Modality A) and structured clinical inputs (Modality B). These components are combined into an early fusion framework that combines features from both modalities and performs the last classification based on a learned boundary decision.

The image data pipeline, as modality A, processes medical image data in the form of mammographic images. The raw images undergo an initial preprocessing step of normalization, resizing, and cropping. Data augmentation can also be applied to improve training diversity. Feature extraction is carried out by a custom CNN or by a pretrained image encoder such as EfficientNet or ResNet, based on the experimental setup. The extracted features from the image are textural and spatial features relevant to the detection of benign and malignant tissue, which are sent to the fusion layer for blending with other modality inputs.

In the clinical data pipeline, as modality B, clinical structured features such as mean radius, texture, perimeter, and other diagnostic measurements are fed into a separate branch of the system. The numerical features are converted using traditional dense layers to produce relevant feature embeddings. Clinical data for certain test cases also pass through a large language model like BioBERT or ClinicalBERT for certain test cases. In these, clinical information is formatted as text-like inputs in the form of medical records that pass through the processing of the LLM to yield deep semantic embeddings. It allows the system to gauge generated text as an alternative or a supplement as a modality.

The WDBC, WBC, and BCSC datasets consist of clinical data, and they serve as the foundational benchmark, where comparisons with more complex models are conducted afterwards to evaluate performance. Different classifiers include LR, XGBoost model, and MLP (benchmark).

Once all extracted features from every modality-specific branch are represented as feature vectors, they are concatenated in the fusion layer. Early fusion

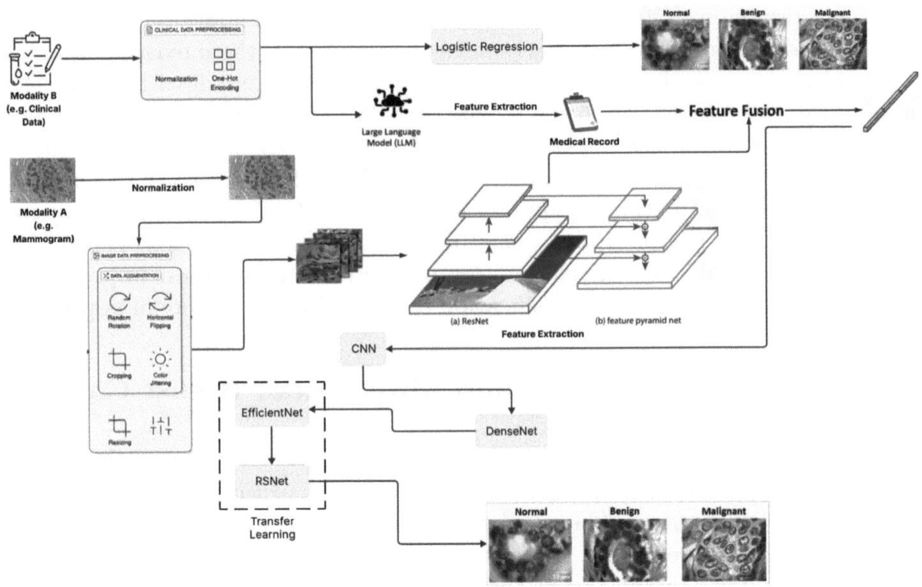

Fig. 2. The overall architecture of the preliminary framework supports both unimodal and multimodal breast cancer detection

enables the system to learn inter-modality relationships as well as take advantage of the complementary nature of the inputs. The fused representation is fed into a final dense layer or a shallow classifier, i.e., LR, which emits a probability of malignancy with a sigmoid activation function.

This architecture integrates many diverse data pipelines into a single classification framework. Each component serves some specific function in supporting clinical data representation and image content processing. One of the strengths of architecture is that it is flexible. It accommodates unimodal setups and multimodal pairs (e.g., image + clinical, clinical + LLM). The flexibility is particularly appropriate for end-to-end experimentation and in real-world healthcare settings. It allows for rigorous testing of the usefulness of TL. The extensible and modular structure is a great starting point for both experimental study and future clinical deployment environments.

4 Implementation

In the image processing pipeline, a ResNet-18 backbone pretrained on ImageNet, fine-tuned on the mammogram dataset without the ensemble self-attention layers, was implemented. For the pre-trained ResNet-18 backbone, ImageNet weights were loaded for all convolutional layers to leverage general visual features. The original 1,000-unit fully connected layer was replaced with a 3-unit layer (Normal, Benign, Malignant) followed by SoftMax activation. L2 weight decay (2×10^{-4}) and gradient clipping (max_norm = 1.0) were applied to stabilize training. AdamW with AMSGrad enabled weight updates. No additional attention or fusion modules were included in this baseline, allowing us to isolate the pure image-based performance before adding other data modalities or network enhancements.

The high-level block-level overview of the image processing pipeline (see Fig. 3) provides a clear picture of how feature extraction, dimensionality reduction, and classification are arranged. Block A represents the ResNet-18 backbone, which takes a $3 \times 224 \times 224$ RGB mammogram input and outputs a $512 \times 7 \times 7$ feature map. Block B is a 1×1 convolutional "FPN Conv" layer ($512 \rightarrow 256$) with BatchNorm, ReLU, and Dropout (p = 0.5). Block C applies AdaptiveAvgPool2d to yield a $1 \times 1 \times 256$ tensor and then flattens it to a 256-dimensional vector. Finally, block D is the two-layer classification head (Dense $256 \rightarrow 128 \rightarrow$ LayerNorm $\rightarrow$ ReLU $\rightarrow$ Dropout, then Dense $128 \rightarrow 3 \rightarrow$ SoftMax).

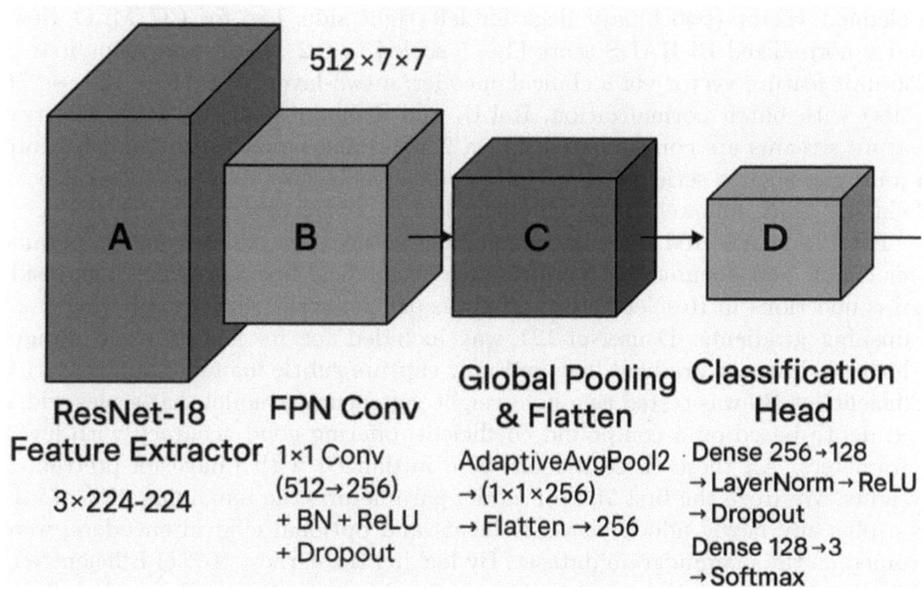

Fig. 3. The high-level block-level overview of the image processing pipeline for the feature extraction, dimensionality reduction, and classification

The extracted feature map first passed through a 7×7 Conv2d layer ($3 \rightarrow 64$ channels, stride 2, padding 3), followed by BatchNorm2d and ReLU, then a 3×3 MaxPool (stride 2, padding 1). This is followed by the residual stages (Layer 1 - Layer 4), each consisting of two BasicBlocks and channel expansions from $64 \rightarrow 128 \rightarrow 256 \rightarrow 512$. The resulting $512 \times 7 \times 7$ feature map is reduced to a 1×1 Conv2d ($512 \rightarrow 256$) with BatchNorm, ReLU, and Dropout (p = 0.5), then pooled via AdaptiveAvgPool2d to a $1 \times 1 \times 256$ tensor. After flattening to a 256-dimensional vector, the classifier head applies Linear ($256 \rightarrow 128$) $\rightarrow$ LayerNorm $\rightarrow$ ReLu $\rightarrow$ Dropout ($p = 0.5$), and finally Linear ($128 \rightarrow 3$) to produce class logits (followed by SoftMax at inference). This architecture balances pretrained feature extraction with a lightweight custom head for breast cancer classification.

The multimodal pipeline combines image features extracted by a pretrained CNN backbone with clinical metadata processed by a small, fully connected network. The input mammogram (224×224 RGB) passed through a ResNet-50 model (with its original final classification layer removed). After loading ImageNet-pretrained weights, replace its 1,000-unit fully connected layer with an identity mapping. The model's structure remains: an initial 7×7 convolution + batch norm + ReLU + max-pool, followed by the residual blocks (layer 1 - layer 4) with bottleneck units. The global average pooling that follows layer 4 produces a 2,048-dimensional feature vector. A configurable freeze ratio (70%) locks the earliest convolutional layers to preserve general visual representations while fine-tuning deeper layers. Meanwhile, a clinical-encoder network takes a 5-element vector (two binary flags for left/right side, two for CC/MLO view, and a normalized BI-RADS score ($1 - 5$ scaled to $0.2 - 1.0$)) and maps it to a 256-unit feature vector via a clinical encoder: a two-layer MLP ($5 \rightarrow 128 \rightarrow 256$ units) with batch normalization, ReLU, and dropout (0.4 then 0.3). The two feature streams are concatenated into a 2,304-dimensional feature and fed into a joint classifier: a series of three fully connected layers ($1024 \rightarrow 512 \rightarrow 256 \rightarrow$ 3 classes), each followed by batch norm, ReLU, and dropout (0.5, 0.4, 0.3).

For TL, we selected three backbone architectures that balance depth, parameter count, and documented performance on medical imaging tasks. The residual connections in ResNet-50 can train deeper networks without suffering from vanishing gradients. DenseNet-121 was included for its feature-reuse design, which can improve gradient flow and may capture subtle mammogram patterns. EfficientNet-B0 was tested as a lightweight yet accurate model that scales width and depth based on a compound coefficient, offering good accuracy with fewer parameters. All these three models were initialized with ImageNet-pretrained weights. We froze the first 70% of model parameters; the remaining 30% of layers, plus any newly added classifier heads and optional clinical encoders, were trained on the mammogram dataset. By locking the earliest 50% of EfficientNet-B0's weights, the model retains general visual representations learned on ImageNet.

5 Evaluation and Experimental Analysis

WDBC, WBC, BCSC, and AISSLAB were used in this study. The first three are structured unimodal, consisting of structured clinical data, while AISSLAB consists of real-world mammography images with associated metadata. Comparing the performance of the framework with unimodal datasets and the performance of the same framework with multimodal datasets is crucial.

For the unimodal experiments, each of these datasets requires its own preprocessing pipeline subject to its data modality. The testing and experimenting

Table 1. The hyperparameter settings for classifiers in unimodal experiments

Classifier	Hyperparameters
LR	Max_iter = 1000 (or 10000 for larger datasets), standard L2 regularization
MLP (WDBC only)	Two hidden layers of sizes 32 and 16, ReLU, Adam optimizer, training was done up to a maximum of 1,000 iterations
XGBoost (BCSC only)	100 boosting rounds and logloss objective, default learning rate and regularization

with these models were carried out using CPU-optimized libraries. All unimodal datasets were balanced with the SMOTE technique. Once the preprocessing and balancing steps were taken, the dataset was split into training and testing subsets using an 80/20 stratified split. Model-specific hyperparameters are presented in Table 1. The LR, RF, and XGBoost models were trained using a single-pass fitting routine. RF and XGBoost models were trained with 100 estimators/rounds, respectively.

For all the TL and multimodal experiments, 20% of the mammogram images for a final test set are reserved, 20% for validation during training, and the remaining 60% for model fitting. Additionally, in the unimodal mammogram images classification experiments, a stratified 5-fold cross-validation on the 80% training and validation split was employed. In each fold, 80% of that subset was used for training and 20% for validation, with the held-out 20% test set used only for the final evaluation. A consistent set of core hyperparameters is maintained as follows:

Learning rate: 3×10^{-4} for AdamW (TL) or AdamW with AMSGrad in the unimodal runs.
Batch size: 32 images per mini-batch.
Weight decay: 1×10^{-4} in TL experiments, 2×10^{-4} in the image-only baselines to prevent overfitting.
Optimizer settings: AdamW with gradient clipping to stabilize training gradients.
Scheduler: CosineAnnealingWarmRestarts ($T_0 = 10$) for the TL pipeline, and ReduceLROnPlateau (factor = 0.5, patience = 3 epochs) for the pure image models.
Augmentation: Random horizontal flips (50%), vertical flips (20%), rotations ($\pm 20^{c}irc$), affine translations ($\pm$15%), and color jitter in the transfer-learning scripts.

Early stopping was triggered when the validation F1 failed to improve for 10 consecutive epochs, preventing unnecessary over-training. All models were trained for up to 100 epochs.

The performance summary of unimodal models across all datasets is presented in Table 2. Across the three datasets tested, unimodal classifiers proved

Table 2. The performance summary of unimodal models across all datasets

Dataset	Model	Accuracy	F1 score	AUC	Precision	Recall
WDBC	LR	0.958	0.96	0.99	0.97	0.95
WDBC	MLP	0.993	0.99	1	0.99	1
WBC	LR	0.9663	0.9651	0.99	0.9651	0.9651
BCSC	LR	0.69	0.7	0.7545	0.69	0.7
BCSC	RF	0.92	0.92	0.9764	0.92	0.92
BCSC	XGBoost	0.9	0.9	0.9731	0.91	0.89

to have strong predictive ability. The MLP on the WDBC dataset achieved the highest performance metric amongst all models tested, with an accuracy of 0.993 and an F1 score of 0.99. LR showed promising results for both the WDBC and the WBC dataset, with F1 scores of 0.96 and 0.9651, respectively. This is confirmation that simple linear models produce promising results when applied to well-preprocessed structured data. Following promising results, the perfor-

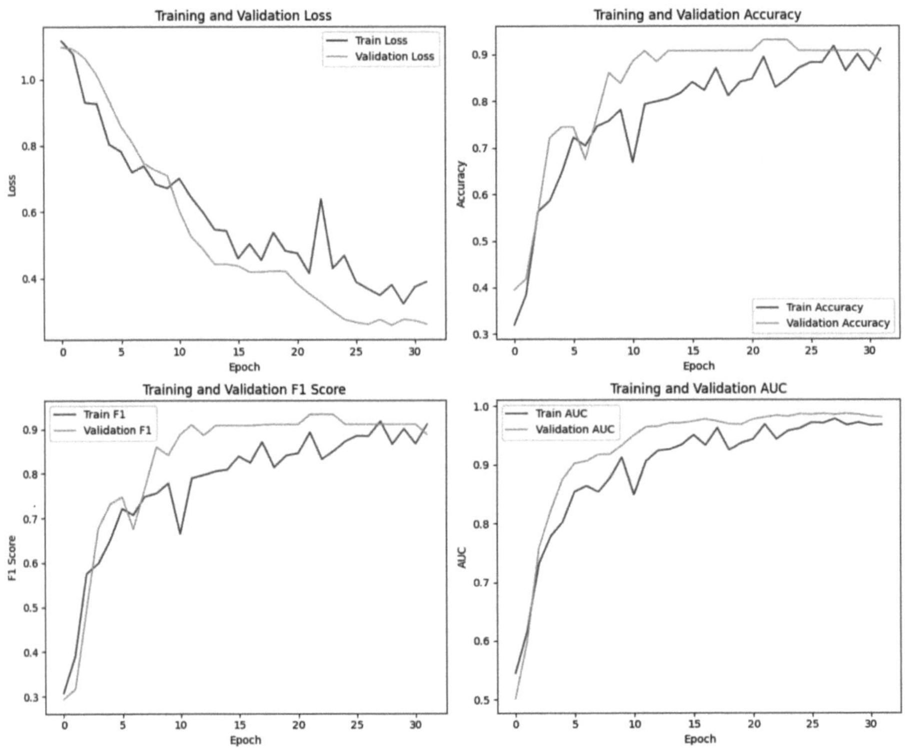

Fig. 4. TL training and validation results

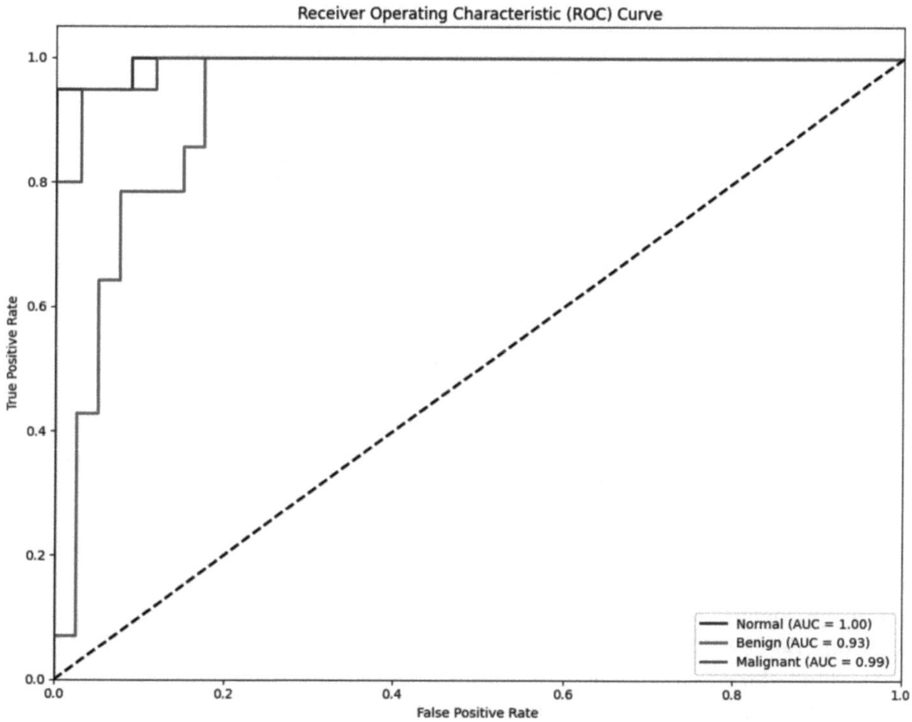

Fig. 5. The ROC curve for the TL model

mance of the BCSC dataset varied significantly. The ensemble methods (RF and XGBoost) outperformed LR greatly, demonstrating their capacity to model complex, non-linear relationships in high-dimensional clinical datasets. RF achieved a perfect balance between precision and recall (both 0.92). XGBoost performed best in precision (0.91); however, it lagged slightly in recall (0.89).

The best TL configuration (ResNet-50 backbone, 70% frozen, 30% fine-tuned, with clinical metadata) is evaluated. Figure 4 shows the training and validation loss, F1-score, and AUC across 30 epochs for the TL model. The loss curves drop rapidly in the first 10 epochs (from 1.1 to 0.5), then taper off smoothly to around 0.3 by epoch 30. The small gap between training and validation loss indicates that the freeze-and-fine-tune strategy (freezing 70% of layers) plus weight decay (1×10^{-4}) and dropout (0.5) effectively controlled overfitting. Accuracy and F1 climb in parallel, reaching 0.92 by epoch 20 and peaking at 0.95. AUC improves from 0.75 to 0.97, showing that the pretrained features adapt well to mammogram classification with only modest fine-tuning.

Figure 5 shows class-wise ROC curves and AUC values for the TL model. The "normal" class achieves an AUC of 0.98, indicating near-perfect discrimination. The "malignant" class follows closely at 0.97, reflecting the model's strong ability to detect cancer patterns. The "benign" class yields a slightly lower AUC of 0.93, matching its lower recall in the confusion matrix. These high AUCs confirm that fine-tuning deeper layers while retaining general ImageNet features produces a classifier that can distinguish subtle mammogram differences, especially between healthy and malignant tissues.

6 Conclusion and Future Work

Breast cancer remains one of the leading causes of cancer-related deaths worldwide. This study investigated the value of multimodal learning for breast cancer classification by integrating mammographic images along with structured clinical metadata. The aim was to investigate whether various multimodal setups would enhance diagnostic accuracy and adaptability. This is particularly important in real-world settings, where having both structured clinical data and mammograms/scans of a patient is the most reliable method for achieving the most accurate early diagnosis.

Unimodal models (MLP neural networks, RF, LR, XGBoost) across three different datasets (WDBC, WBC, BCSC) served as a foundational baseline, providing a benchmark for evaluating the effectiveness of more complex multimodal approaches. The text-only model, which uses only BI-RADS scores and view/side flags, achieves the highest overall performance (F1 = 96.2%, AUC = 98%). This shows that structured clinical metadata alone carries very strong signals for distinguishing between normal, benign, and malignant cases.

Combining mammogram images with metadata via TL on EfficientNet-B0 yields the second-best performance (F1 score = 81%, AUC = 92%). The pre-trained visual features, once fine-tuned and balanced, complement the metadata to boost discrimination, especially improving malignant detection while still benefiting from clear metadata patterns. The EnhancedFusionNet variant, which was designed to fuse image features with a small LLM analysis at inference, lags all other methods (F1 score = 77.3%, AUC = 85%). This result suggests that the model did not add meaningful information beyond the basic metadata that had already been encoded. A more specialized or larger LLM, trained directly in radiology reports, may be needed to see real gains. The results of these experiments revealed that various complex multimodal configurations can enhance the performance of the models; however, a lot of the time, the unimodal models (baselines) trained on just clinical metadata outperformed the multimodal models. For instance, across all experiments conducted (unimodal and multimodal), the MLP classifier trained on the WDBC dataset achieved the highest results (an F1 score of 99.13% and an accuracy of 99.04%). These metrics outperformed every multimodal setup. The baseline model for the multimodal experiments was only successful in outperforming the LR performed on the BCSC dataset (achieved an accuracy of 69%). For small clinical datasets, low-dimensional metadata can sometimes match or even exceed image-only methods, and the best performance often comes from smartly combining both data and model innovations.

This study was successful in confirming the utility of multimodal architectures for breast cancer detection. Ultimately, in real-world cases, having both mammograms and structured patient data is considered ideal for healthcare professionals; the more data that is available, the better it is for diagnosing. However, the results of this study prove that in the context of ML, working with multimodal models does not always generate superior outcomes compared to simpler unimodal approaches.

Future work could explore larger and more diverse datasets and domain-specific LLMs trained in real radiology text interpreting model outputs. Potential applications include generating natural language explanations of predictions or enhancing interpretability through text-image alignment.

Acknowledgments. We would like to thank the Department of Computer Science at Munster Technological University for supporting this research.

References

1. Breast Cancer (2024). https://www.who.int/news-room/fact-sheets/detail/breast-cancer/. Accessed 10 May 2025
2. Society, I.C.: Breast Cancer (2025). https://www.cancer.ie/cancer-information-and-support/cancer-types/breast-cancer. Accessed 10 May 2025
3. Marie Keating Foundation: Breast Cancer: What you should know (2025). https://mariekeating.ie/cancer-information/breast-cancer/breast-cancer-what-you-should-know/
4. Jafari, A.: Machine-learning methods in detecting breast cancer and related therapeutic issues: a review. Comput. Methods Biomech. Biomed. Eng. Imaging Visual. **12**(1) (2024). https://doi.org/10.1080/21681163.2023.2299093
5. Yue, W., Wang, Z., Chen, H., Payne, A., Liu, X.: Machine learning with applications in breast cancer diagnosis and prognosis. Designs **2**(2), 13 (2018). https://doi.org/10.3390/designs2020013
6. Islam, M.M., Haque, M.R., Iqbal, H., Hasan, M.M., Hasan, M., Kabir, M.N.: Breast cancer prediction: a comparative study using machine learning techniques. SN Comput. Sci. **1**(5), 1–14 (2020). https://doi.org/10.1007/s42979-020-00305-w
7. Islam, T., Sheakh, M.A., Tahosin, M.S., et al.: Predictive modeling for breast cancer classification in the context of Bangladeshi patients by use of machine learning approach with explainable AI. Sci. Rep. **14**, 8487 (2024). https://doi.org/10.1038/s41598-024-57740-5
8. Younas, H.I., Bukhari, S., Bukhari, F., Aslam, N., Badar, H.M.S., Kajla, N.I.: An efficient methodology for the classification of invasive ductal carcinoma using transfer learning. J. Comput. Biomed. Inf. **4**(01), 236–250 (2022)
9. Hakeem, A., Hussain, I., Qadri, S., Mehdi, M., Kajla, N.I., Akbar, A.J.: Computer vision based solanum lycopersicon leaf disease detection using transfer learning. J. Comput. Biomed. Inf. **6**(02), 402–412 (2024)
10. Sharma, S., Aggarwal, A., Choudhury, T.: Breast cancer detection using machine learning algorithms. In: 2018 International Conference on Computational Techniques, Electronics and Mechanical Systems (CTEMS), pp. 114–118 (2018). https://doi.org/10.1109/CTEMS.2018.8769187
11. Jasti VDP, Zamani, A. S., Arumugam, K., Naved, M., Pallathadka, H., Sammy, F., Kaliyaperumal, K.: Computational technique based on machine learning and image processing for medical image analysis of breast cancer diagnosis. Secur. Commun. Netw. 1–7 (2022). https://doi.org/10.1155/2022/1918379

12. Nemade, V., Fegade, V.: Machine learning techniques for breast cancer prediction. Procedia Comput. Sci. **218**, 1314–1320 (2023). https://doi.org/10.1016/j.procs.2023.01.110
13. Zafar, U.B., Hamza, H., Kajla, N.I., Badar, H.M.S., Nawaz, S.A., Siddique, M.N.: Advancing glioblastoma diagnosis through innovative deep learning image analysis in histopathology. J. Comput. Biomed. Inf. (2024)
14. Subramanian, V., Syeda-Mahmood, T., Do, M.N.: Multimodal fusion using sparse CCA for breast cancer survival prediction. In: IEEE International Symposium on Biomedical Imaging, p. 14291432 (2021)
15. Ben Rabah, C., Sattar, A., Ibrahim, A., Serag, A.: A multimodal deep learning model for the classification of breast cancer subtypes. Diagnostics **15**(8), 995 (2025). https://doi.org/10.3390/diagnostics15080995
16. Mondol, R.K., Millar, E.K., Sowmya, A., Meijering, E.: MM-SurvNet: deep learning-based survival risk stratification in breast cancer through multimodal data fusion. ArXiv arXiv:2402.11788 (2024)
17. Londhe, G.V., Birchha, V., Naresh Kumar Reddy, V., Veena, S., Srilakshmi, A., Karthigaikumar, P.: Multimodal deep learning for breast cancer detection: integrating transfer learning with clinical data for enhanced accuracy. In: Proceedings of the 3rd International Conference on Optimization Techniques in the Field of Engineering (ICOFE-2024) (2024). Available at SSRN: https://ssrn.com/abstract=5066783, https://doi.org/10.2139/ssrn.5066783

Intelligent Agents in Engineering Systems

Transfer Learning for Low Pressure Plasma Etching Rate Prediction

Eneri Boniakou[1(✉)], Tatiana Boura[2,3,4], Alex Kondi[5], Vassilios Constantoudis[5], and George Kokkoris[1]

[1] School of Chemical Engineering, National Technical University of Athens, Zografou, Greece
eneriboniakou@mail.ntua.gr
[2] Institute of Machine Learning and Neural Computation, TU Graz, Graz, Austria
[3] Graz Center for Machine Learning, Graz, Austria
[4] Institute of Informatics and Telecommunications, National Centre of Scientific Research "Demokritos", Ag. Paraskevi, Greece
[5] Institute of Nanoscience and Nanotechnology, National Centre of Scientific Research "Demokritos", Ag. Paraskevi, Greece

Abstract. Low-pressure inductively coupled plasmas are essential in semiconductor manufacturing for etching and thin-film processing. Their ability to generate high plasma densities and independently control the ion energy increases etching rate and enables anisotropic etching. However, accurately predicting spatially varying etching rates remains a challenge. Physics-based models (PBMs) capture the complex plasma dynamics but are computationally expensive, requiring the solution of large systems of partial differential equations. Optimization tasks demand repeated PBM evaluations, compounding the cost. Additionally, plasma processes vary significantly in gas chemistry, reactor configuration, and operating conditions, necessitating separate models for each case. In this work, we frame the optimization process as a multi-agent system, where neural predictive agents replace PBMs for specific chemistries and conditions, collectively supporting plasma process optimization. We propose a machine learning approach using these predictive agents trained to predict etching rates for two distinct chemistries. We apply transfer learning techniques to fine-tune a model trained on one chemistry for use with another, thereby reducing the amount of PBM-generated training data required. While transfer learning is well-established in fields such as computer vision and natural language processing, its use in physics-based tasks, where the problem structure differs significantly, remains limited. Our results demonstrate that high accuracy can be achieved through transfer learning with limited data, and that an "inverse" layer-freezing strategy, tuning early layers while freezing later ones, yields superior performance for our task, achieving R^2 scores of 0.93 and 0.65, compared to alternative layer-freezing strategies, which reach at best 0.50 and 0.00, respectively.

Keywords: Transfer learning · Plasma processing · Machine learning · Chemical engineering

© The Author(s), under exclusive license to Springer Nature Switzerland AG 2026
K. Mason et al. (Eds.): IASE 2025, CCIS 2976, pp. 85–101, 2026.
https://doi.org/10.1007/978-3-032-28263-7_6

1 Introduction

Plasma, often referred to as the fourth state of matter, is a collection of charged and neutral particles and is on average electrically neutral. Technological plasmas are usually low-pressure plasmas and are widely used in semiconductor manufacturing for the etching of substrates and deposition of thin films; about 40–45% of the steps to manufacture semiconductor devices use plasma in one form or another [7]. Plasma etching takes place in low pressure plasma reactors: through the application of electromagnetic fields, reactive species are produced in the reactor bulk which are then transferred on the substrate and cause etching; a common and critical specification of etching is the etching uniformity on the substrate.

Despite the progress in plasma technology, accurate predictions of local etching rates on the substrate remain challenging due to the complex interplay of electromagnetic fields, plasma species transport, and gas/surface chemistry. The accurate prediction of local etching rates is crucial for the optimization of the plasma etching process, and enables a comprehensive understanding of how the operating parameters (power, pressure, gas composition at the inlet, and substrate temperature) affect the etching rate.

Multidimensional physics-based models (PBMs) capture these coupled phenomena but entail a high computational cost, as they solve extensive sets of partial differential equations for mass, energy, and momentum balances alongside Maxwells equations. As a result, the problem becomes significantly nonlinear, which can elevate the complexity of optimization process–an aspect of paramount importance for industrial applications. An optimization algorithm requires multiple iterations of the computationally expensive PBM. Therefore, a model capable of making predictions orders of magnitude faster can significantly facilitate optimization algorithms [27].

A natural approach to reducing computational time is the development of data-driven surrogate models that map reactor operating conditions directly to substrate-scale etching rates. Artificial intelligence (AI), and specifically Machine Learning (ML), techniques can be employed as the foundation for such models. By framing these ML models as predictive agents, they can be integrated into broader decision-making multi-agent systems for adaptive process control. These agents are trained on data obtained from PBMs and serve as predictors of etching rates.

Various gases or their mixtures are employed for plasma etching of different substrates, with each process having specific operating conditions, reactor configurations, and chemical kinetics. Due to this diversity, it becomes essential to develop PBMs for each individual process. Consequently, data must be generated for each process (defined by the substrate, the operating conditions, reactor configurations, and chemical kinetics) to train the ML algorithms.

One solution to reduce the excessive amount of data required to feed different algorithms for different chemistries is to train a predictive agent on one chemistry and then apply techniques to transfer knowledge to another plasma chemistry. This process is particularly feasible with neural network (NN) architectures, as they serve as function approximators. In function space terminology, transfer learning is advantageous because the learned function in the source chemistry already approximates a meaningful mapping from inputs to outputs. If the target chemistry shares underlying physical principles, the optimal target function lies in a region of the function space close to the source function, reducing the search space for optimization. So, instead of learning the target function from scratch, the model starts from a well-informed prior. To further enhance transfer learning, layer freezing plays a crucial role by constraining optimization to a relevant subspace [24]. This strategy can be interpreted as structuring the agent's internal knowledge layers to balance general adaptability with task-specific precision. Moreover, transfer learning in NNs is an established methodology and has demonstrated significant success in other domains such as computer vision and natural language processing [12,22].

However, this existing knowledge about transfer learning is mostly derived from these former applications which differ fundamentally from those in physical sciences. A primary example of this misalignment is the common recommendation to freeze the initial layers during fine-tuning, under the assumption that they capture general knowledge. In physics-related tasks, this assumption may not hold; in our application domain, generalizable knowledge may be more likely stored in the later layers, since the shallow layers are more likely to adapt to the operating conditions of each case and its governing chemical kinetics.

Our approach conceptualizes neural networks as predictive agents capable of adapting their predictive capabilities across different plasma chemistries through transfer learning. Specifically, we explore transfer learning techniques using NNs to predict the local etching rates of a Poly(methyl methacrylate) (PMMA) blanket sample under two distinct chemistries: oxygen (O_2) and argon (Ar) plasma within a low pressure inductively coupled plasma reactor. Based on our findings, we propose a methodology suitable for practical application. Our goal is twofold: first, to achieve successful transfer learning, i.e., minimizing the amount of data needed for fine-tuning, and second, to employ the most appropriate techniques that consider the structure and the nature of our task.

The structure of our paper is as follows: In Sect. 2, we provide a brief literature review on the application of AI in plasma modeling, followed by a discussion on the use of transfer learning in physics-based problems. Section 3 presents an overview of the PBMs used for etching rate prediction. In Sect. 4, we conduct and discuss transfer learning experiments on predictive agents trained with the two different chemistries. Finally, Sect. 5 provides our conclusion.

2 Background

2.1 AI in Low Pressure Plasmas

ML-based agents have emerged as a powerful tool for addressing the complexity of partial differential equations that describe unit operations and reaction kinetics in chemical engineering applications [27]. These predictive agents can operate autonomously within larger optimization or control frameworks, interacting with other agents within the system to adapt reactor conditions in real time. Its integration into the field of plasma technology has shown great promise across a wide range of use cases, particularly in optimizing reactor performance and modeling complex plasma processes.

For instance, physics informed neural networks (PINNs) [20] have been applied to optimize various plasma reactor parameters, such as the Peclet number and temperature [8]. PINN-based agents have also been effectively used to model plasma dynamics, providing insights into electric potentials and fields [14]. Additionally, Desai et al. [4] model the complex interaction of electromagnetic waves with plasma through 2D simulations using convolutional neural networks (CNNs) and capture spatiotemporal discharge dynamics [4].

Multilayer perceptrons (MLPs) have been used as predictive agents to model chemical reaction kinetics in pulsed discharge simulations [29,32] and have also been proposed to predict the electron velocity distribution function, with their accuracy validated for two reference gases [10]. In this case, the MLPs were used to solve the Boltzmann equation while accounting for ionization and electron attachment processes. MLPs have also been employed to predict elastic momentum cross-sections of electrons in gases by leveraging experimental data from established databases [16]. In plasma chemistry modeling, MLPs have shown the potential to replace traditional kinetic models, offering significant improvements in computational efficiency. In fact, Pan et al. [19] reported a remarkable reduction in computational cost from 30 h to just a few seconds when using MLPs to simulate plasma discharges.

Intelligent agents have also been applied in modeling sputtering phenomena, where a surrogate model was used to simulate argon ion bombardment of a composite target. The MLP model successfully mapped the energy distribution of sputtered particles to the angular ion energy distribution [11]. Additionally, intelligent agents that employ MLPs, decision trees and support vector regression models have been utilized to analyze current and voltage data in plasma reactors, enabling real-time tuning of the electrical matching network [9].

Furthermore, agents that utilize recurrent neural networks have been employed in the model predictive control of etching processes [30,31]. In this context, neural networks were used to predict discharge dynamics during Silicon etching with high speed and accuracy.

2.2 Transfer Learning for Physics Problems

The efficiency of Deep Learning is strongly linked to the existence of large annotated datasets, which are in general expensive and hard to obtain. To address this problem, transfer learning techniques, such as fine-tuning and freezing, have been widely used, particularly in fields like computer vision and natural language processing. In fine-tuning, the idea is that when two problems are similar–often assessed using transferability metrics [1,2,5,18,28]–a network can first be trained on a source task and then further trained on the target task using the pre-trained model. By doing so, the target model not only converges more quickly but also requires less data, as the source model already possesses a strong knowledge base that simply needs to be fine-tuned for the specific task. In systems where neural networks operate as intelligent agents, this capability enables faster adaptation to new tasks or environments with minimal retraining, improving both efficiency and responsiveness in dynamic settings.

Another key assumption when transferring knowledge from one neural network to another is that general features are learned in certain parts of the network, while task-specific features are modeled in others. This hierarchical feature organization also supports modular agent design, where different components of an intelligent agent specialize in either general knowledge retention or task-specific adaptation. Specifically, the lower layers of the neural network capture low-level concepts, while the layers closer to the output focus on features more directly related to the task. The idea of freezing involves keeping the initial layers unchanged (and possibly gradually unfreeze them), as they have already acquired useful general knowledge, and only retrain the final layers to fine-tune them for the specific task at hand.

In the field of physics, transfer learning is particularly appealing, as physical phenomena can change over time, and transfer learning provides a robust framework for adapting models to slightly out-of-distribution data. This adaptability is particularly valuable for autonomous systems and multi-agent frameworks managing complex physical processes, where continual learning and cross-domain knowledge transfer are essential. Several studies have explored these ideas, either by fine-tuning the neural network [13] or by freezing its shallow layers [3,6,26], following common practices in the field.

However, the effectiveness of freezing layers, as well as the evidence supporting this approach, depends highly on the domain of application. For example, in vision-based transfer learning tasks [21], the input remains consistent between the source and target tasks (i.e., still images), while the output varies. In this case, the assumption that shallow layers contain essential knowledge holds true, as fundamental image processing remains the same regardless of the task. In engineering and physics, however, two seemingly similar problems (as judged by domain experts) may differ in their input, output, or underlying governing laws. This means we cannot automatically assume that all universal knowledge is stored at the beginning of the network. For agents applied to physics-based tasks, this observation suggests the need for flexible knowledge transfer strategies rather than fixed layer-freezing protocols.

This idea has not been widely explored, with one study [25] challenging this established wisdom on the domain of weather and climate modeling. Their findings suggest that when a CNN is used to model turbulent flow, freezing the deepest layers yields the best results. They further support this conclusion with spectral analysis and model-specific evaluations.

3 Physics-Based Model for Low-Pressure Plasma Etching

In this work, we model plasma etching in the Gaseous Electronics Conference (GEC) reference reactor (cf. Figure 1) operating in inductively coupled plasma

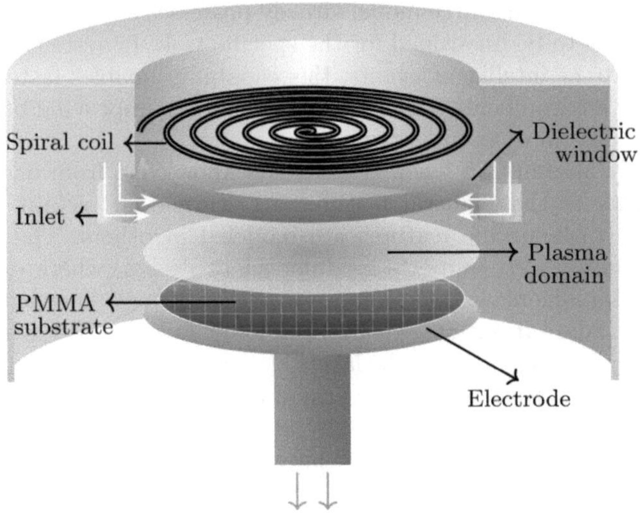

Fig. 1. Plasma discharge in the GEC reactor. The main components in a cross-section of the reactor are shown, including the plasma.

(ICP) mode. The reactor features a five-turn spiral coil positioned above a dielectric window, through which couples with the plasma. The coil is driven by a 13.56 MHz generator. In the computational domain, the gas inlet is positioned below the dielectric window, while the outlet lies beneath the lower electrode, which is placed opposite the window. The energy is transmitted to the electrons inductively by the coil as it does not come in direct contact with the plasma gas.

Given the reactor's axial symmetry, the PBM is solved in a 2D axisymmetric geometry, which significantly reduces computational time compared to a 3D geometry. The PBM includes the mass balances for neutral and charged species and electrons, the electron energy balance, the momentum conservation equation, Ampére's law to determine the magnetic field induced by the coil, and Poisson's equation to compute the electric potential in the plasma reactor [17]. The inputs of the PBM are two: namely the *coil power* and *pressure* and the outputs are the spatial distribution of species densities, electric potential, electron temperature in the reactor, as well as the local etching rates along the wafer radius of the PMMA sample. In our predictive application, we are interested in mapping the inputs to the etching rates; therefore, we use only the latter as the output of the PBM.

This study focuses on etching of blanket samples of PMMA with Ar and O_2 plasmas. The surface etching kinetics are based on the studies of Memos et al. [15] and Spyrides et al. [23]. The O_2 reaction set consists of 156 reactions, whereas the Ar chemistry is simpler, comprising 40 reactions [17]. Additionally, the O_2 PBM includes the formation of six ion species, whereas the Ar plasma considers only a single ion species. The O_2 plasma is governed by an ion-enhanced etching mechanism while the Ar plasma etching proceeds solely through physical sputtering. The previous differences make the O_2 chemistry notably more complex than that of Ar.

For the O_2 model, the operating conditions–pressure and coil power–were varied within the range [20–70 mTorr, 500–2000 W], while for the Ar model, they were varied in the range [10–60 mTorr, 50–500 W]. These differences reflect the distinct gas chemistries, resulting in different etching rates, as in the O_2 they fall within the range of 0.1 to 0.4 μm/min and for Ar the range is from 0.0001 to 0.089 μm/min.

Regarding the PBMs' run-time, a single run of the Ar PBM takes 45 min, whereas for the O_2, it takes approximately twice as long.[1]

4 Experiments and Results

4.1 Experimental Setup

The PBMs for the O_2 and Ar plasmas presented in Sect. 3 are executed iteratively to generate the data that feeds the neural networks (predictive agents). The total number of data points for each plasma chemistry is 4000. We follow a 60-20-20% split for the training, validation, and testing sets. However, the testing set is selected from a subset of the data by choosing a distinct subspace within the grid formed by the inputs, for fair evaluation. Specifically, for the O_2 data, we selected inference points where the power falls within the range [1340 W, 2000 W] and the pressure falls within [46 mTorr, 70 mTorr]. For the Ar data, the selected power range is [330 W, 500 W], and the pressure range is [39 mTorr, 60 mTorr].

The neural networks' architecture is a simple feedforward model performing regression, with two inputs and ten outputs corresponding to different etching rate values across the wafer. To study the effects of transfer learning, we trained two separate models, each using data from the two different chemistries. To ensure a fair comparison for transfer learning, we selected models with the same capacity through hyperparameter tuning. Specifically, both the O_2 network, denoted as M_{O_2}, and the Ar network, M_{Ar}, consist of three layers, amounting to 580 trainable parameters per network. The optimal training hyperparameters were determined via grid search. For the Ar network, we used a batch size of 4, a learning rate of 2.94×10^{-3}, a weight decay of 1.74×10^{-6}. For the O_2 network, we used a batch size of 32, a learning rate of 9.62×10^{-4}, and a weight decay of 2.06×10^{-5}. Both models were trained for 700 epochs with a patience of 20.

To train the models, we implemented a weighted mean squared error loss. This approach was chosen because as we move radially outward on the wafer, there is a significant reduction in the etching rate. In fact, the etching rate at the center of the wafer can differ by an order of magnitude compared to the edge, due to higher species densities at the center. This difference in the magnitude made it difficult for the network to learn the latter five outputs when they all

[1] The hardware configuration regarding the PBMs throughout this work consists of an AMD Ryzen Threadripper 3960X 24-Core Processor with 264GB of RAM.

Table 1. Evaluation of the trained models on the respective test sets using Mean Absolute Error (MAE) and the coefficient of determination (R^2). M_{O_2} denotes the neural network trained on O_2 plasma data, while M_{Ar} denotes the model trained on Ar plasma data. The selected metrics represent the macro mean of all outputs. To interpret the MAE (μm/min), consider the output value ranges for each model.

Metric	**Baseline Models**	
	M_{Ar}	M_{O_2}
MAE	3.43×10^{-5}	1.64×10^{-2}
$\mathbf{R^2}$	0.926	0.819

contributed with an equal percentage to the loss. To address this, we weighted their corresponding loss terms to ensure a balanced contribution to the loss function.

To ensure good generalization, we applied early stopping and L2 as regularization techniques. We opted for Min-Max normalization of the data, with the respective global values. This approach ensures fair normalization for the testing sets and provides unique, consistent scaling for both networks, which is essential for transfer learning in regression tasks.

To perform transfer learning, we conducted a series of standard experiments, including fine-tuning pre-trained networks with a reduced learning rate, exploring catastrophic forgetting, freezing layers and combining different freezing strategies, gradually unfreezing previously frozen layers, and investigating training with a reduced amount of data. We should note that, in all experiments, we used only two test sets: one for the Ar plasma and another for the O_2 plasma.[2]

4.2 Results and Discussion

Initially, we train the M_{O_2} and M_{Ar} to obtain a baseline performance on the initial task, which is presented in Table 1. The performance is evaluated using a subset of the standard regression metrics: Mean Absolute Error (MAE) and the coefficient of determination (R^2). We select these metrics because MAE is a more intuitive measure for domain experts, while R^2 is generally more interpretable. In our problem, MAE is measured in μm/min. R^2 has an upper bound of 1, which represents the ideal outcome, while lower values indicate progressively worse performance. As a global measure of model performance, we compute the mean of these metrics across all ten outputs. For a more detailed evaluation, wherever we believe is necessary, we assess each output individually.

[2] The code and data are available at https://zenodo.org/records/15343316.

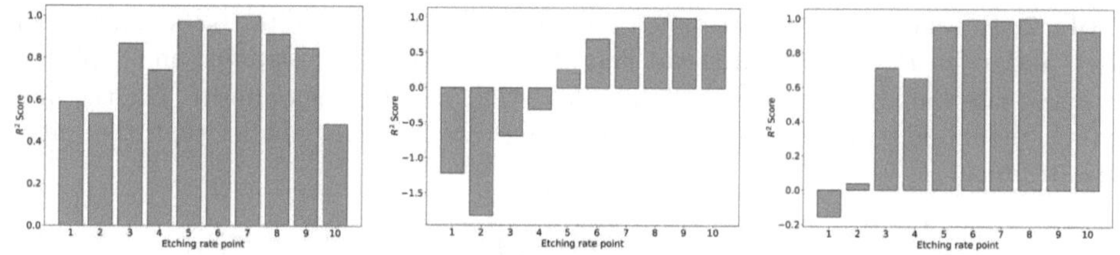

(a) M_{O_2} (b) M_{Ar}^{O2} fine-tuned (c) M_{Ar}^{O2} frozen last layer

Fig. 2. Per-output R^2 on the O_2 test set performing inference on: (a) the initial model trained on the O_2 data, (b) the fine-tuned pre-trained Ar model on the O_2 data, and (c) the pre-trained Ar model with the last layer frozen fine-tuned on the O_2 data.

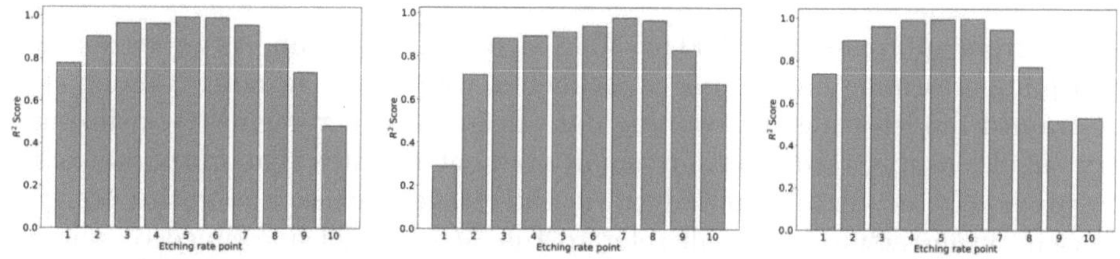

(a) M_{Ar} (b) $M_{O_2}^{Ar}$ fine-tuned (c) $M_{O_2}^{Ar}$ frozen last layer

Fig. 3. Per-output R^2 on the Ar test set performing inference on: (a) the initial model trained on the Ar data, (b) the fine-tuned pre-trained O_2 model on the Ar data, and (c) the pre-trained O_2 model with the last layer frozen fine-tuned on the Ar data.

Interpreting Table 1 alongside Fig. 2a, we see that the etching rates of the O_2 plasma are harder to predict, as on the first two and the last etching rate point the individual R^2 score falls below 0.6 on the test set. On the other hand, the Ar plasma is a more suitable task for this configuration, as indicated by Fig. 3a, since only the last column is not predicted with high accuracy. This is further validated by the higher complexity of the O_2 chemistry, as described in Sect. 3.

However, we observe that although the average metric is high, the performance is not evenly distributed, and some etching rate points are more difficult to predict. So, apart from the motivation that we need transfer learning to obtain networks trained on multiple chemistries without the excessive use of expensive PBMs, it is interesting to explore transfer learning from another perspective: whether it can help improve the prediction of etching rates that are not well captured.

Fine-Tuning. Before investigating the idea of transfer learning and searching for the most suitable techniques to yield the best results, we performed a simple test where we conducted inference on the pre-trained model using the opposite test set, i.e., feeding the Ar test data to M_{O_2} and vice versa. As expected, the results were poor in both scenarios, showing negative individual R^2 values

Table 2. Evaluation of the fine-tuned models on the respective test sets using mean MAE and mean R^2. $M_{O_2}^{Ar}$ denotes the neural network that is pre-trained on O_2 plasma data and fine-tuned on Ar plasma data, while $M_{Ar}^{O_2}$ denotes the opposite. For the fine-tuning, the models were trained with a learning rate 20 times smaller than the initial (pre-)training.

Metric	**Fine-tuned Models**	
	$M_{O_2}^{Ar}$	$M_{Ar}^{O_2}$
MAE	7.28×10^{-3}	3.92×10^{-2}
$\mathbf{R^2}$	0.822	−0.152

ranging from approximately -500 to -10 in each output column. As a consequence, we began performing fine-tuning experiments, where we continued training each pre-trained model using the training data from the other chemistry, utilizing a very small learning rate to ensure small gradient updates that would not erase the previously acquired knowledge. At this stage, we use the entire training datasets. From now on, we will refer to the M_{O_2} model fine-tuned on the Ar data as $M_{O_2}^{Ar}$, and to the M_{Ar} model fine-tuned on the O_2 data as $M_{Ar}^{O_2}$.

In Table 2, we observe that fine-tuning performs fairly well for the Ar plasma, as there is only a minor drop of 0.1 in the R^2 value compared to the baseline model (Table 1). Regarding the individual R^2 scores, in Fig. 3b, we observe that although the mean score of the fine-tuned model is lower than the baseline, the fine-tuned model performs worse than the baseline in the first two etching rate points but outperforms it in the last four. Thus, transfer learning here adds value by "fixing" certain outputs. For the O_2 plasma, the results are overall suboptimal. As illustrated in Fig. 2b, the first half of the outputs are heavily under-predicted, with only the last individual output outperforming the baseline.

Note that fine-tuning is a transfer learning technique that, in essence, serves as a form of pre-training, providing a more informed initialization of the network. However, it does not account for the structural differences between the source and target problems in transfer learning. Instead, it relies on the assumption that this initialization will lead to better local minima in the loss function during training. Therefore, it is not surprising that its results were suboptimal. In physics-related problems, and particularly in our case, there is an underlying structure where certain components benefit more from fine-tuning than others; specifically, it makes more sense to adjust the initial layers related to the input chemistry, rather than the later layers that capture the behavior of the PBM simulation.

Catastrophic Forgetting. After fine-tuning, we also tested the fine-tuned models to see whether they suffered from *catastrophic forgetting*, meaning whether they had forgotten the initial task they were trained on, i.e., the initial chemistry. We observed that both agents, in both cases, forgot their initial task. However, we note that for the domain of interest, this is a desired outcome, as opposed to what the general wisdom in machine learning states. We want to ensure that the

Table 3. Evaluation of multiple models where transfer learning was applied by freezing certain layers. We present results from four experiments: (a) Freezing the first input layer of the pre-trained model, (b) Freezing the last output layer of the pre-trained model, (c) Freezing the middle and last layers of the pre-trained model, and (d) Gradually unfreezing the last layer of the pre-trained model after 100 epochs.

Metric	Freeze First Layer		Freeze Last Layer		Freeze Middle & Last Layer		Gradually Unfreeze Last Layer	
	$M_{O_2}^{Ar}$	$M_{Ar}^{O_2}$	$M_{O_2}^{Ar}$	$M_{Ar}^{O_2}$	$M_{O_2}^{Ar}$	$M_{Ar}^{O_2}$	$M_{O_2}^{Ar}$	$M_{Ar}^{O_2}$
MAE	1.42×10^{-2}	4.65×10^{-2}	3.75×10^{-3}	$\mathbf{1.66 \times 10^{-2}}$	1.70×10^{-2}	5.00×10^{-2}	$\mathbf{4.56 \times 10^{-3}}$	3.21×10^{-2}
$\mathbf{R^2}$	0.504	−0.908	0.911	**0.646**	0.294	−0.003	**0.932**	0.575

agents forget the old chemistry entirely, as it is irrelevant to the new task. This ties back to the statement we made in Sect. 2.2, that the status quo for transfer learning in the machine learning domain should be revisited and evaluated across different domains to derive more fine-grained principles.

Freezing. Knowing the structure of the task at hand, we suspect that the initial layers adapt to the operating conditions of each case and its governing chemical reaction mechanism, while the deeper layers account for the solution of the PDEs and the surface mechanism, which are simulated similarly in both PBMs.

We perform a series of preliminary experiments to validate our assumptions, which involve (a) freezing individual layers, (b) freezing all combinations of layers, and (c) gradually unfreezing the frozen layers. We only present the results of relevant experiments here.

As presented in Table 3, our assumptions are experimentally validated, as the "inverse" transfer learning by freezing the last layer produced the best results out of all freezing experiments and also outperformed fine-tuning (cf. Table 2). In our experiments, we demonstrate that the last layer holds the most crucial knowledge: freezing it along with the middle layer, led to significantly worse results. Regarding the Ar plasma data, this transfer learning technique yielded results comparable to the initial M_{Ar} model trained on Ar data. The tuned model predicts the middle outputs almost perfectly but performs poorly on the edge outputs, as illustrated in Fig. 3c.

For the O_2 plasma data, this method underperformed the performance of the initial model. However, as seen in Fig. 2c, the frozen model made errors different to the initial model (depicted in Fig. 2a). The last six columns are predicted almost perfectly, but the model underperforms in the first two. This, again, may suggest that transfer learning could be a useful approach to improve the performance of certain components.

Lastly, it is interesting to note that gradually unfreezing the last layer (the configuration that performed best) did not globally yield better results but only slightly improved the performance of $M_{O_2}^{Ar}$. This further validates our initial assumption that the last layer holds universal knowledge shared by the two chemistries. The best approach is to adjust the model to this knowledge by training the earlier layers, rather than attempting to modify the acquired knowledge.

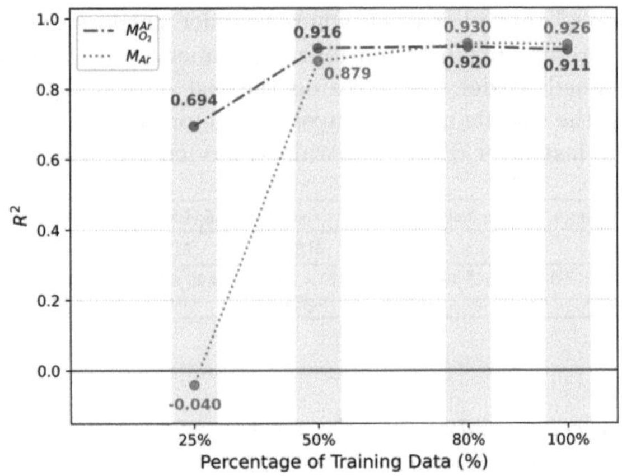

Fig. 4. Mean R^2 progression for models trained on a percentage of the training data. We evaluate the baseline model M_{Ar}, as well as the pre-trained model with the last layer frozen ($M_{O_2}^{Ar}$).

Using Less Data. Let us revisit the initial question: Can we achieve good performance without the need to generate expensive data from multiple chemistries using PBMs? This question is relevant only in the context of Ar plasma chemistry, as transfer learning yields results comparable to the baseline. In contrast, for O_2 the performance is inferior, making it unlikely to improve with less data.

So, we conduct experiments that progressively use a percentage of the Ar training data for transfer learning. Each time, the subset of the training data is selected randomly while maintaining the original distribution, but the test set remains the same as the one used in all the experiments. To perform these experiments, we select the pre-trained model with the last layer frozen, as it performed best on average among all transfer learning methods and combinations in the validation set. Additionally, for comparison, we train the baseline model with the reduced amount of data.

Regarding $M_{O_2}^{Ar}$ and M_{Ar} in Fig. 4, we observe that for up to 50% of the data, the loss of precision is not significant in both the baseline and the tuned models. However, when dropping below 50%, the decline in accuracy becomes abrupt, falling below zero in the baseline model, whereas in the fine-tuned model, it becomes steeper but remains relatively high. This suggests that with just 25% of the initial data–rather than half–transfer learning can still achieve results comparable to the original training.

So, in this transfer learning scenario, the computational savings from avoiding the need for PBMs to compute all combinations is significant, as illustrated in Fig. 5. Specifically, by using only 25% of the Ar, we gain approximately 50 h of parallel PBM runs compared to the initial 100%, and about 25 h less than the 50% usage.

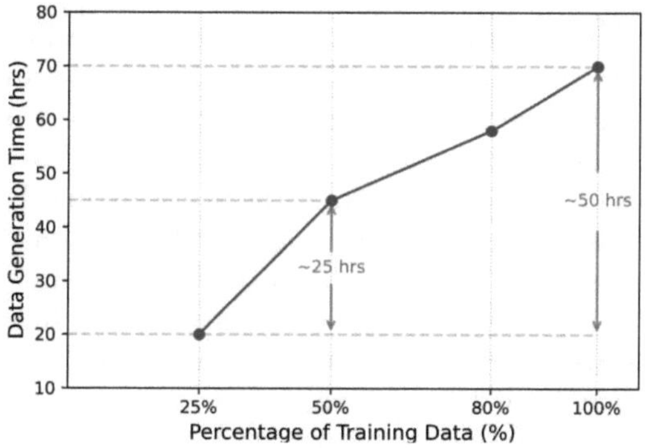

Fig. 5. Hours required for the *Ar* plasma PBM to generate the percentages of data, using parallelization. Each parallel run utilized 8 cores.

Transfer Learning: Source Agent. As mentioned in Sect. 3, the O_2 plasma chemistry is more complex than the *Ar* chemistry. This difference in complexity is also reflected in the performance of the neural networks in the experiments. However, it is further evident in the divergence of success in the transfer learning experiments, as the majority of transfer learning scenarios are feasible for the *Ar* chemistry, but none are for the O_2. The above can be interpreted as follows: when the source chemistry is more complex than the target, transfer learning is possible and is expected to perform well. On the other hand, transitioning from a simpler chemistry to a more complex one is challenging, and the only feasible scenario would be if the model forgets the *Ar* chemistry and only learns the O_2, which, of course, is a futile approach.

In the context of these predictive agents integrated into a multi-agent system utilized for plasma process modeling, this behavior highlights a practical limitation: an agent trained on simpler plasma chemistries lacks the representational capacity to handle the richer dynamics of more complex chemistries without substantial retraining. The knowledge acquired in the simpler domain does not sufficiently cover the broader state-action space required for accurate predictions in the more complex scenario and this will lead as well to poor collaboration within the system.

5 Conclusions and Future Work

In this work, we applied neural networks as predictive agents within a broader multi-agent system designed to optimize plasma processing tasks. These predictive agents replace expensive PBMs and were investigated for their ability to migrate knowledge from one plasma chemistry to another via transfer learning, aiming to reduce the need for generating costly data when training models for two different plasma chemistries.

We performed experiments involving fine-tuning a pre-trained model, freezing its layers, and training with progressively less data for O_2 and *Ar* plasmas. Each predictive agent in this system focuses on modeling a specific plasma chemistry while sharing knowledge with other agents to enhance system-wide efficiency.

For the simpler Ar chemistry, transferring knowledge from the more complex O_2 plasma was entirely successful. This transfer produced predictions with a small error margin (R^2 of 0.93), making these agents highly reliable and suitable for production use within the multi-agent plasma control framework. We also found that using only a quarter of the initial training data yielded an acceptable error margin, with an R^2 of 0.70, which does not preclude the method's use when large datasets are unavailable. On the other hand, for the O_2 chemistry, transferring knowledge from the pre-trained Ar agent was not feasible. In all experiments, the fine-tuned models did not surpass the performance of the baseline model. The best R^2 achieved with transfer learning in this case was 0.65, whereas the baseline model reached an R^2 of 0.82. This indicates a directional asymmetry in knowledge transfer capability among predictive agents, which has implications for designing multi-agent systems where agents of differing complexity levels must interact. In both cases, we found that an "inverse" freezing approach–freezing the deeper layers of the network–performed best, yielding R^2 scores of 0.93 and 0.65, compared to alternative layer-freezing strategies, which achieved at best 0.50 and 0.00, respectively. This contrasts with the conventional wisdom in machine learning, which typically favors freezing the initial layers. This observation suggests that, within plasma processing systems, predictive agents should be fine-tuned with strategies adapted to the structure and nature of the specific task rather than relying on generic transfer learning conventions.

Regarding the chemistries involved, we believe that the unsuccessful transfer of knowledge from Ar to O_2 is due to the inherent complexity of each chemistry. The latter is far more complex than the former, and using Ar as a source underestimates the complexity of O_2. From a multi-agent system perspective, this implies that agents trained on simpler chemistries contribute best as supporting elements rather than as primary sources of knowledge for more complex agents. In the future, we aim to perform experiments on other single-element plasmas and evaluate this claim by considering all source-target combinations and accounting for their complexity within the agent system architecture.

However, of greater importance are the mixtures of gases, rather than single-element plasmas. These mixtures are much more complex in terms of the reactions, requiring intricate PBMs to generate their data and resulting in significantly enhanced run times. In such cases, the exploration of transfer learning among predictive agents becomes even more critical for maintaining system-wide efficiency. Therefore, we aim to explore the concept of transfer learning in these scenarios and, specifically, how to transition from simple chemistries to more complex ones using coordinated agent strategies. Not only will this approach enable us to draw conclusions about physics-inspired transfer learning tasks, where input features share commonalities across different chemistries, but it will also allow us to conduct a series of experiments with deeper neural networks of larger capacity. This, in turn, will help us develop a more refined theory on layer freezing tailored to our specific scenario and better inform the design of predictive agents within plasma processing multi-agent systems. Additionally, we intend to investigate transfer learning in ensembles of networks acting as collaborative predictive agents for the mixture of gases. This approach involves transitioning from multiple agents trained on simple elements to a unified agent model that predicts the behavior of their mixtures. By combining the strengths of individual agents, we hope to improve the predictive power and generalization of the overall system while also requiring less data for training.

Acknowledgments. This work is supported by the project "Enhancing plasma etching efficiency, repeatability, and environmental footprint via AI-based modeling and optimization (plasmAI)" of the program "AI-Aware Pathways to Sustainable Semiconductor Process and Manufacturing Technologies", Intel Corporation & Merck KGaA.

Disclosure of Interests. Authors have no competing interests.

References

1. Bao, Y.: An information-theoretic approach to transferability in task transfer learning. In: 2019 IEEE International Conference on Image Processing (ICIP), pp. 2309–2313 (2019). https://doi.org/10.1109/ICIP.2019.8803726
2. Bolya, D., Mittapalli, R., Hoffman, J.: Scalable diverse model selection for accessible transfer learning. Adv. Neural. Inf. Process. Syst. **34**, 19301–19312 (2021)
3. Chakraborty, S.: Transfer learning based multi-fidelity physics informed deep neural network. J. Comput. Phys. **426**, 109942 (2021). https://doi.org/10.1016/j.jcp.2020.109942
4. Desai, M., Ghosh, P., Kumar, A., Chaudhury, B.: Deep-learning architecture-based approach for 2-D-simulation of microwave plasma interaction. IEEE Trans. Microw. Theory Tech. **70**, 1–10 (2022). https://doi.org/10.1109/TMTT.2022.3217138
5. Dwivedi, K., Roig, G.: Representation similarity analysis for efficient task taxonomy & transfer learning. In: 2019 IEEE/CVF Conference on Computer Vision and Pattern Recognition (CVPR), pp. 12379–12388 (2019). https://doi.org/10.1109/CVPR.2019.01267.
6. Goswami, S., Anitescu, C., Chakraborty, S., Rabczuk, T.: Transfer learning enhanced physics informed neural network for phase-field modeling of fracture. Theor. Appl. Fract. Mech. **106**, 102447 (2020). https://doi.org/10.1016/j.tafmec.2019.102447
7. Graves, D.B., et al.: Science challenges and research opportunities for plasma applications in microelectronics. J. Vacuum Sci. Technol. B **42**, 042202 (2024). https://doi.org/10.1116/6.0003531
8. Hou, Q., Du, H., Sun, Z., Wang, J., Wang, X., Wei, J.: PINN-CDR: a neural network-based simulation tool for convection-diffusion-reaction systems. Int. J. Intell. Syst. (2023). https://doi.org/10.1155/2023/2973249
9. Hussain, S., Lary, D., Hara, K., Bera, K., Rauf, S., Goeckner, M.: Case study in machine learning for predicting moderate pressure plasma behavior. J. Vac. Sci. Technol., A **42**, 043001 (2024). https://doi.org/10.1116/6.0003434
10. Kawaguchi, S., Takahashi, K., Ohkama, H., Satoh, K.: Deep learning for solving the Boltzmann equation of electrons in weakly ionized plasma. Plasma Sources Sci. Technol. **29**, 025021 (2020). https://doi.org/10.1088/1361-6595/ab6074

11. Krüger, F., Gergs, T., Trieschmann, J.: Machine learning plasma-surface interface for coupling sputtering and gas-phase transport simulations. Plasma Sources Sci. Technol. **28**, 035002 (2019). https://doi.org/10.1088/1361-6595/ab0246
12. Li, X., et al.: Transfer learning in computer vision tasks: remember where you come from. Image Vis. Comput. **93**, 103853 (2020). https://doi.org/10.1016/j.imavis.2019.103853
13. Liu, Y., Liu, W., Yan, X., Guo, S., Zhang, C.: Adaptive transfer learning for PINN. J. Comput. Phys. **490**, 112291 (2023). https://doi.org/10.1016/j.jcp.2023.112291
14. Mathews, A., Francisquez, M., Hughes, J.W., Hatch, D.R., Zhu, B., Rogers, B.N.: Uncovering turbulent plasma dynamics via deep learning from partial observations. Phys. Rev. E **104**(2), 025205 (2021). https://doi.org/10.1103/PhysRevE.104.025205
15. Memos, G., Lidorikis, E., Gogolides, E., Kokkoris, G.: A hybrid modeling framework for the investigation of surface roughening of polymers during oxygen plasma etching. J. Phys. D Appl. Phys. **54**, 175205 (2021). https://doi.org/10.1088/1361-6463/abdb0b
16. Morgan, W.L.: Use of numerical optimization algorithms to obtain cross sections from electron swarm data. Phys. Rev. A **44**(3), 1677–1681 (1991). https://doi.org/10.1103/PhysRevA.44.1677
17. Mouchtouris, S., Kokkoris, G.: A hybrid model for low pressure inductively coupled plasmas combining a fluid model for electrons with a plasma-potential-dependent energy distribution and a fluid-Monte Carlo model for ions. Plasma Sources Sci. Technol. **25**, 025007 (2016). https://doi.org/10.1088/0963-0252/25/2/025007
18. Nguyen, C.V., Hassner, T., Seeger, M., Archambeau, C.: LEEP: a new measure to evaluate transferability of learned representations. In: Proceedings of the 37th International Conference on Machine Learning (2020). https://doi.org/10.5555/3524938.3525614
19. Pan, J., Liu, Y., Zhang, S., Hu, X., Liu, Y., Shao, T.: Deep learning-assisted pulsed discharge plasma catalysis modeling. Energy Convers. Manage. **277**, 116620 (2023). https://doi.org/10.1016/j.enconman.2022.116620
20. Raissi, M., Perdikaris, P., Karniadakis, G.E.: Physics-informed neural networks: a deep learning framework for solving forward and inverse problems involving nonlinear partial differential equations. J. Comput. Phys. **378**, 686–707 (2019). https://doi.org/10.1016/j.jcp.2018.10.045
21. Hakeem, A., Hussain, I., Qadri, S., Mehdi, M., Kajla, N.I., Akbar, A.J.: Computer vision based solanum lycopersicon leaf disease detection using transfer learning. J. Comput. Biomed. Inf. **6**(02), 402–412 (2024)
22. Ruder, S., Peters, M.E., Swayamdipta, S., Wolf, T.: Transfer learning in natural language processing. In: Proceedings of the 2019 Conference of the North American Chapter of the Association for Computational Linguistics: Tutorials, pp. 15–18 (2019). https://doi.org/10.18653/v1/N19-5004.
23. Spyrides, S., Alencastro, F., Guimaraes, E., Bastian, F., Simão, R.: Mechanism of oxygen and argon low pressure plasma etching on polyethylene (UHMWPE). Surf. Coat. Technol. **378**, 124990 (2019). https://doi.org/10.1016/j.surfcoat.2019.124990
24. Younas, H.I., Bukhari, S., Bukhari, F., Aslam, N., Badar, H.M.S., Kajla, N.I.: An efficient methodology for the classification of invasive ductal carcinoma using transfer learning. J. Comput. Biomed. Inf. **4**(01), 236–250 (2022)
25. Subel, A., Guan, Y., Chattopadhyay, A., Hassanzadeh, P.: Explaining the physics of transfer learning in data-driven turbulence modeling. PNAS Nexus **2**(3), pgad015 (2023). https://doi.org/10.1093/pnasnexus/pgad015

26. Tang, H., Liao, Y., Yang, H., Xie, L.: A transfer learning-physics informed neural network (TL-PINN) for vortex-induced vibration. Ocean Eng. **266**, 113101 (2022). https://doi.org/10.1016/j.oceaneng.2022.113101
27. Toscano, J.D., et al.: From PINNs to PIKANs: recent advances in physics-informed machine learning. In: Machine Learning for Computational Science and Engineering, vol. 1, no. 1, p. 15 (2025). https://doi.org/10.1007/s44379-025-00015-1.
28. Tran, A., Nguyen, C., Hassner, T.: Transferability and hardness of supervised classification tasks. In: 2019 IEEE/CVF International Conference on Computer Vision (ICCV), pp. 1395–1405 (2019). https://doi.org/10.1109/ICCV.2019.00148.
29. Wang, X.C., Zhang, Y.T.: Modeling of discharge characteristics and plasma chemistry in atmospheric CO_2 pulsed plasmas employing deep neural network. J. Appl. Phys. **133**, 143301 (2023). https://doi.org/10.1063/5.0143741
30. Xiao, T., Ni, D.: Multiscale modeling and recurrent neural network based optimization of a plasma etch process. Processes **9**, 151 (2021). https://doi.org/10.3390/pr9010151
31. Xiao, T., Wu, Z., Christofides, P., Armaou, A., Ni, D.: Recurrent neural-network-based model predictive control of a plasma etch process. Ind. Eng. Chem. Res. **61**, 638–652 (2021). https://doi.org/10.1021/acs.iecr.1c04251
32. Zhang, Y.T., Gao, S.H., Ai, F.: Efficient numerical simulation of atmospheric pulsed discharges by introducing deep learning. Front. Phys. **11**, 1125548 (2023). https://doi.org/10.3389/fphy.2023.1125548

Author Index

© The Editor(s) (if applicable) and The Author(s), under exclusive license
to Springer Nature Switzerland AG 2026

K. Mason et al. (Eds.): IASE 2025, CCIS 2976, p. 103, 2026.
https://doi.org/10.1007/978-3-032-28263-7

MIX
Papier aus verantwortungsvollen Quellen
Paper from responsible sources
FSC® C105338

If you have any concerns about our products,
you can contact us on
ProductSafety@springernature.com

In case Publisher is established outside the EU,
the EU authorized representative is:
Springer Nature Customer Service Center GmbH
Europaplatz 3, 69115 Heidelberg, Germany

Printed by Libri Plureos GmbH
in Hamburg, Germany